Welcome to *My Math* – your very own math book! You can write in it – in fact, you are encouraged to write, draw, circle, explain, and color as you explore the exciting world of mathematics. Let's get started. Grab a pencil and finish each sentence.

My name is _____.

My favorite color is _____.

My favorite hobby or sport is _____.

My favorite TV program or video game is
_____.

My favorite class is _____.

mhmymath.com

Copyright © 2018 McGraw-Hill Education

All rights reserved. No part of this publication may be reproduced or distributed in any form or by any means, or stored in a database or retrieval system, without the prior written consent of McGraw-Hill Education, including, but not limited to, network storage or transmission, or broadcast for distance learning.

STEM McGraw-Hill is committed to providing instructional materials in Science, Technology, Engineering, and Mathematics (STEM) that give all students a solid foundation, one that prepares them for college and careers in the 21st century.

Send all inquiries to:
McGraw-Hill Education
8787 Orion Place
Columbus, OH 43240

ISBN: 978-0-07-905764-8 (**Volume 2**)
MHID: 0-07-905764-0

Printed in the United States of America.

6 7 8 9 10 11 QSX 23 22 21 20 19

Understanding by Design® is a registered trademark of the Association for Supervision and Curriculum Development ("ASCD").

McGraw-Hill My Math

Grade 4 • Volume 2

Authors:
Carter • Cuevas • Day • Malloy
Altieri • Balka • Gonsalves • Grace • Krulik • Molix-Bailey
Moseley • Mowry • Myren • Price • Reynosa • Santa Cruz
Silbey • Vielhaber

GO digital ▶ connectED.mcgraw-hill.com

▶ Log In

1 Go to **connectED.mcgraw-hill.com**.

2 Log in using your username and password.

3 Click on the Student Edition icon to open the Student Center.

▶ Go to the Student Center

4 Click on Menu, then click on the **Resources** tab to see all of your online resources arranged by chapter and lesson.

5 Click on the **eToolkit** in the Lesson Resources section to open a library of eTools and virtual manipulatives.

6 Look here to find any assignments or messages from your teacher.

7 Click on the **eBook** to open your online Student Edition.

Username: _____

Password: _____

▶ Explore the eBook!

8 Click the **speaker icon** at the top of the eBook pages to hear the page read aloud to you.

 Practice and review your Vocabulary.

 Animations and videos allow you to explore mathematical topics.

 Explore concepts with eTools and virtual manipulatives.

 Personal Tutors are online virtual teachers that walk you through the steps of the lesson problems.

More resources can be found by clicking the icons at the bottom of the eBook's pages.

9 eHelp helps you complete your homework.

 Explore these fun digital activities to practice what you learned in the classroom.

 Worksheets are PDFs for Math at Home, Problem of the Day, and Fluency Practice.

Contents in Brief
Organized by Domain

Number and Operations in Base Ten

Chapter 1	Place Value
Chapter 2	Add and Subtract Whole Numbers
Chapter 3	Understand Multiplication and Division
Chapter 4	Multiply with One-Digit Numbers
Chapter 5	Multiply with Two-Digit Numbers
Chapter 6	Divide by a One-Digit Number

Operations and Algebraic Thinking

Chapter 7	Patterns and Sequences

Number and Operations — Fractions

Chapter 8	Fractions
Chapter 9	Operations with Fractions
Chapter 10	Fractions and Decimals

Measurement and Data

Chapter 11	Customary Measurement
Chapter 12	Metric Measurement
Chapter 13	Perimeter and Area

Geometry

Chapter 14	Geometry

connectED.mcgraw-hill.com

Chapter 1: Place Value

Number and Operations in Base Ten

ESSENTIAL QUESTION
How does place value help represent the value of numbers?

Getting Started

- My Chapter Project 2
- Am I Ready? 3
- My Math Words 4
- My Vocabulary Cards 5
- My Foldable FOLDABLES 9

Lessons and Homework

- **Lesson 1** Place Value 11
- **Lesson 2** Read and Write Multi-Digit Numbers 17
- **Lesson 3** Compare Numbers 23
- **Lesson 4** Order Numbers 29
- Check My Progress 35
- **Lesson 5** Use Place Value to Round 37
- **Lesson 6** Problem-Solving Investigation: Use the Four-Step Plan 43

Wrap Up

- My Chapter Review 49
- Reflect 52
- Performance Task 52PT1

Look for this! Click online and you can watch videos that will help you learn the lessons.

connectED.mcgraw-hill.com

vii

Number and Operations in Base Ten

ESSENTIAL QUESTION
What strategies can I use to add or subtract?

Chapter 2 Add and Subtract Whole Numbers

Getting Started

My Chapter Project	54
Am I Ready?	55
My Math Words	56
My Vocabulary Cards	57
My Foldable FOLDABLES	59

Lessons and Homework

Lesson 1	Addition Properties and Subtraction Rules	61
Lesson 2	Addition and Subtraction Patterns	67
Lesson 3	Add and Subtract Mentally	73
Lesson 4	Estimate Sums and Differences	79
Check My Progress		85
Lesson 5	Add Whole Numbers	87
Lesson 6	Subtract Whole Numbers	93
Lesson 7	Subtract Across Zeros	99
Check My Progress		105
Lesson 8	Problem-Solving Investigation: Draw a Diagram	107
Lesson 9	Solve Multi-Step Word Problems	113

Wrap Up

Fluency Practice	119
My Chapter Review	121
Reflect	124
Performance Task	124PT1

connectED.mcgraw-hill.com

Chapter 3
Understand Multiplication and Division

Number and Operations in Base Ten

ESSENTIAL QUESTION
How are multiplication and division related?

Getting Started

My Chapter Project	126
Am I Ready?	127
My Math Words	128
My Vocabulary Cards	129
My Foldable FOLDABLES	133

Lessons and Homework

Lesson 1 Relate Multiplication and Division 135
Lesson 2 Relate Division and Subtraction 141
Lesson 3 Multiplication as Comparison 147
Lesson 4 Compare to Solve Problems 153
Check My Progress 159
Lesson 5 Multiplication Properties and Division Rules 161
Lesson 6 The Associative Property of Multiplication 167
Lesson 7 Factors and Multiples 173
Lesson 8 Problem-Solving Investigation: Reasonable Answers 179

Wrap Up

My Chapter Review 185
Reflect 188
Performance Task 188PT1

Look for this! Click online and you can get more help while doing your homework.

connectED.mcgraw-hill.com

Number and Operations in Base Ten

ESSENTIAL QUESTION
How can I communicate multiplication?

Chapter 4
Multiply with One-Digit Numbers

Getting Started

- My Chapter Project **190**
- Am I Ready? **191**
- My Math Words **192**
- My Vocabulary Cards **193**
- My Foldable FOLDABLES **195**

Lessons and Homework

- **Lesson 1** Multiples of 10, 100, and 1,000 .... **197**
- **Lesson 2** Round to Estimate Products **203**
- **Lesson 3** Hands On: Use Place Value to Multiply **209**
- **Lesson 4** Hands On: Use Models to Multiply **215**
- Check My Progress **221**
- **Lesson 5** Multiply by a Two-Digit Number **223**
- **Lesson 6** Hands On: Model Regrouping **229**
- **Lesson 7** The Distributive Property **235**
- **Lesson 8** Multiply with Regrouping **241**
- **Lesson 9** Multiply by a Multi-Digit Number **247**
- Check My Progress **253**
- **Lesson 10** Problem-Solving Investigation: Estimate or Exact Answer **255**
- **Lesson 11** Multiply Across Zeros **261**

Wrap Up

- My Chapter Review **267**
- Reflect **270**
- Performance Task **270PT1**

connectED.mcgraw-hill.com

Chapter 5
Multiply with Two-Digit Numbers

Number and Operations in Base Ten

ESSENTIAL QUESTION
How can I multiply by a two-digit number?

Getting Started

My Chapter Project	272
Am I Ready?	273
My Math Words	274
My Vocabulary Cards	275
My Foldable FOLDABLES	277

Lessons and Homework

Lesson 1	Multiply by Tens	279
Lesson 2	Estimate Products	285
Check My Progress		291
Lesson 3	Hands On: Use the Distributive Property to Multiply	293
Lesson 4	Multiply by a Two-Digit Number	299
Lesson 5	Solve Multi-Step Word Problems	305
Lesson 6	Problem-Solving Investigation: Make a Table	311

Wrap Up

My Chapter Review	317
Reflect	320
Performance Task	320PT1

Look for this! Click online and you can find tools that will help you explore concepts.

connectED.mcgraw-hill.com

Number and Operations in Base Ten

ESSENTIAL QUESTION
How does division affect numbers?

Chapter 6: Divide by a One-Digit Number

Getting Started

My Chapter Project 322
Am I Ready? 323
My Math Words 324
My Vocabulary Cards 325
My Foldable FOLDABLES 327

Lessons and Homework

Lesson 1 Divide Multiples of 10, 100, and 1,000 329
Lesson 2 Estimate Quotients 335
Lesson 3 Hands On: Use Place Value to Divide 341
Lesson 4 Problem-Solving Investigation: Make a Model 347
Lesson 5 Divide with Remainders 353
Lesson 6 Interpret Remainders 359
Check My Progress 365
Lesson 7 Place the First Digit 367
Lesson 8 Hands On: Distributive Property and Partial Quotients 373
Lesson 9 Divide Greater Numbers 379
Check My Progress 385
Lesson 10 Quotients with Zeros 387
Lesson 11 Solve Multi-Step Word Problems 393

Wrap Up

Fluency Practice 399
My Chapter Review 401
Reflect 404
Performance Task 404PT1

connectED.mcgraw-hill.com

Chapter 7: Patterns and Sequences

Operations and Algebraic Thinking

ESSENTIAL QUESTION
How are patterns used in mathematics?

Getting Started

- My Chapter Project 406
- Am I Ready? 407
- My Math Words 408
- My Vocabulary Cards 409
- My Foldable FOLDABLES 411

Lessons and Homework

- **Lesson 1** Nonnumeric Patterns 413
- **Lesson 2** Numeric Patterns 419
- **Lesson 3** Sequences 425
- **Lesson 4** Problem-Solving Investigation: Look for a Pattern 431
- Check My Progress 437
- **Lesson 5** Addition and Subtraction Rules 439
- **Lesson 6** Multiplication and Division Rules 445
- **Lesson 7** Order of Operations 451
- Check My Progress 457
- **Lesson 8** Hands On: Equations with Two Operations 459
- **Lesson 9** Equations with Multiple Operations 465

Wrap Up

- My Chapter Review 471
- Reflect 474
- Performance Task 474PT1

Look for this! Click online and you can watch a teacher solving problems.

connectED.mcgraw-hill.com

Number and Operations - Fractions

? ESSENTIAL QUESTION
How can different fractions name the same amount?

Chapter **8** Fractions

Getting Started

My Chapter Project 476
Am I Ready? 477
My Math Words 478
My Vocabulary Cards 479
My Foldable FOLDABLES 483

Lessons and Homework

Lesson 1 Factors and Multiples 485
Lesson 2 Prime and Composite Numbers 491
Check My Progress 497
Lesson 3 Hands On: Model Equivalent Fractions 499
Lesson 4 Equivalent Fractions 505
Lesson 5 Simplest Form 511
Lesson 6 Compare and Order Fractions 517
Lesson 7 Use Benchmark Fractions to Compare and Order 523
Check My Progress 529
Lesson 8 Problem-Solving Investigation: Use Logical Reasoning 531
Lesson 9 Mixed Numbers 537
Lesson 10 Mixed Numbers and Improper Fractions 543

Wrap Up

My Chapter Review 549
Reflect 552
Performance Task 552PT1

connectED.mcgraw-hill.com

Chapter 9 Operations with Fractions

Number and Operations – Fractions

ESSENTIAL QUESTION
How can I use operations to model real-world fractions?

Getting Started

My Chapter Project	554
Am I Ready?	555
My Math Words	556
My Vocabulary Cards	557
My Foldable **FOLDABLES**	559

Lessons and Homework

Lesson 1	Hands On: Use Models to Add Like Fractions	561
Lesson 2	Add Like Fractions	567
Lesson 3	Hands On: Use Models to Subtract Like Fractions	573
Lesson 4	Subtract Like Fractions	579
Check My Progress		585
Lesson 5	Problem-Solving Investigation: Work Backward	587
Lesson 6	Add Mixed Numbers	593
Lesson 7	Subtract Mixed Numbers	599
Check My Progress		605
Lesson 8	Hands On: Model Fractions and Multiplication	607
Lesson 9	Multiply Fractions by Whole Numbers	613

Wrap Up

My Chapter Review	619
Reflect	622
Performance Task	622PT1

Look for this! Click online and you can find activities to help build your vocabulary.

connectED.mcgraw-hill.com

Number and Operations — Fractions

ESSENTIAL QUESTION
How are fractions and decimals related?

Chapter 10 Fractions and Decimals

Getting Started

My Chapter Project 624
Am I Ready? 625
My Math Words 626
My Vocabulary Cards 627
My Foldable FOLDABLES 629

Lessons and Homework

Lesson 1 Hands On: Place Value Through Tenths and Hundredths 631
Lesson 2 Tenths 637
Lesson 3 Hundredths 643
Check My Progress 649
Lesson 4 Hands On: Model Decimals and Fractions 651
Lesson 5 Decimals and Fractions 657
Lesson 6 Use Place Value and Models to Add 663
Lesson 7 Compare and Order Decimals 669
Lesson 8 Problem-Solving Investigation: Extra or Missing Information 675

Wrap Up

My Chapter Review 681
Reflect 684
Performance Task 684PT1

xvi

connectED.mcgraw-hill.com

Chapter 11 Customary Measurement

Measurement and Data

ESSENTIAL QUESTION
Why do we convert measurements?

Getting Started

My Chapter Project	686
Am I Ready?	687
My Math Words	688
My Vocabulary Cards	689
My Foldable FOLDABLES	695

Lessons and Homework

Lesson 1	Customary Units of Length	697
Lesson 2	Convert Customary Units of Length	703
Lesson 3	Customary Units of Capacity	709
Lesson 4	Convert Customary Units of Capacity	715
Check My Progress		721
Lesson 5	Customary Units of Weight	723
Lesson 6	Convert Customary Units of Weight	729
Lesson 7	Convert Units of Time	735
Check My Progress		741
Lesson 8	Display Measurement Data in a Line Plot	743
Lesson 9	Solve Measurement Problems	749
Lesson 10	Problem-Solving Investigation: Guess, Check, and Revise	755

Wrap Up

My Chapter Review	761
Reflect	764
Performance Task	764PT1

connectED.mcgraw-hill.com

xvii

Measurement and Data

ESSENTIAL QUESTION
How can conversion of measurements help me solve real-world problems?

Chapter 12 Metric Measurement

Getting Started

My Chapter Project	766
Am I Ready?	767
My Math Words	768
My Vocabulary Cards	769
My Foldable FOLDABLES	773

Lessons and Homework

Lesson 1	Metric Units of Length	775
Lesson 2	Metric Units of Capacity	781
Lesson 3	Metric Units of Mass	787
Check My Progress		793
Lesson 4	Problem-Solving Investigation: Make an Organized List	795
Lesson 5	Convert Metric Units	801
Lesson 6	Solve Measurement Problems	807

Wrap Up

My Chapter Review	813
Reflect	816
Performance Task	816PT1

connectED.mcgraw-hill.com

Chapter 13 Perimeter and Area

Measurement and Data

ESSENTIAL QUESTION
Why is it important to measure perimeter and area?

Getting Started

My Chapter Project 818
Am I Ready? 819
My Math Words 820
My Vocabulary Cards 821
My Foldable **FOLDABLES** 823

Lessons and Homework

Lesson 1 Measure Perimeter 825
Lesson 2 Problem-Solving Investigation:
Solve a Simpler Problem 831
Check My Progress 837
Lesson 3 Hands On: Model Area 839
Lesson 4 Measure Area 845
Lesson 5 Relate Area and Perimeter 851

Wrap Up

My Chapter Review 857
Reflect 860
Performance Task 860PT1

connectED.mcgraw-hill.com

Geometry

ESSENTIAL QUESTION
How are different ideas about geometry connected?

Chapter 14 Geometry

Getting Started

My Chapter Project	862
Am I Ready?	863
My Math Words	864
My Vocabulary Cards	865
My Foldable FOLDABLES	871

Lessons and Homework

Lesson 1	Draw Points, Lines, and Rays	873
Lesson 2	Draw Parallel and Perpendicular Lines	879
Check My Progress		885
Lesson 3	Hands On: Model Angles	887
Lesson 4	Classify Angles	893
Lesson 5	Measure Angles	899
Lesson 6	Draw Angles	905
Lesson 7	Solve Problems with Angles	911
Check My Progress		917
Lesson 8	Triangles	919
Lesson 9	Quadrilaterals	925
Lesson 10	Draw Lines of Symmetry	931
Lesson 11	Problem-Solving Investigation: Make a Model	937

Wrap Up

My Chapter Review	943
Reflect	946
Performance Task	946PT1

Chapter

8 Fractions

ESSENTIAL QUESTION

How can different fractions name the same amount?

Now We're Cooking!

Watch a video!

Name _____

MY Chapter Project

Healthy Recipes

1. Look through the recipes to find 6 measurements that use fractions or mixed numbers. Record the ingredients below.

Ingredient	Measurement

2. Compare the fractions and mixed numbers listed in your table. Write them in order from least to greatest:

_____

_____

3. Trade papers with another group and check each other's work.

4. Choose a recipe that you think is both healthy and tasty.

Recipe: _____

We picked this recipe because:

_____

_____

_____

Name ....

Am I Ready?

Graph each fraction on a number line.

1. $\frac{1}{3}$

2. $\frac{1}{6}$

3. $\frac{3}{8}$

Use the number line to determine whether the two fractions are equivalent. Write yes or no.

4. $\frac{1}{2}$ and $\frac{4}{8}$

   _____

5. $\frac{2}{3}$ and $\frac{5}{6}$

   _____

Write each whole number as a fraction.

6. 5 _____   7. 9 _____

Shade the boxes to show the problems you answered correctly.

Online Content at connectED.mcgraw-hill.com

Name ....

MY Math Words

Review Vocabulary

fourths   halves   is equal to (=)
is greater than (>)   is less than (<)   thirds

Making Connections
Use the review vocabulary to describe how each shape is divided.
Then use symbols to compare the examples.

478 Chapter 8 Fractions

MY Vocabulary Cards

Lesson 8–7

benchmark fractions

Lesson 8–2

composite number

Lesson 8–3

denominator

$\frac{1}{3}$ $\frac{3}{4}$

Lesson 8–3

equivalent fractions

$\frac{2}{4} = \frac{1}{2}$

Lesson 8–1

factor pairs

24

1 and 24 (1 × 24 = 24)
2 and 12 (2 × 12 = 24)
3 and 8 (3 × 8 = 24)
4 and 6 (4 × 6 = 24)

Lesson 8–5

greatest common factor

12	1	2	3	4	6	12		
30	1	2	3	5	6	10	15	30

greatest common factor: 6

Lesson 8–10

improper fraction

$\frac{7}{3}$

Lesson 8–6

least common multiple

2	2	4	6	8	10	12
5	5	10	15	20	25	30

least common multiple: 10

Ideas for Use

- Group 2 or 3 common words. Add a word that is unrelated to the group. Then work with a friend to name the unrelated word.
- Use a blank card to write this chapter's essential question. Use the back of the card to write or draw examples that help you answer the question.

A whole number with more than two factors.
Describe what a composite number is based on the visuals shown on this card.
_____
_____

Common fractions that are used for estimation.
When do you see or use benchmark fractions in your everyday life?
_____
_____

Fractions that have the same value.
What is an antonym, or word with an opposite meaning, for *equivalent*?
_____

The bottom number in a fraction. It tells the total number of equal parts.
Write a tip to help you remember which number is the denominator and which is the numerator.
_____
_____

The greatest of the common factors of each of two or more numbers.
The suffix *-est* is added to adjectives to mean "the most." Use another adjective with this suffix in a sentence.
_____

The two factors that are multiplied to find a product.
Write 3 factor pairs of 36.
_____
_____

The least multiple greater than 0 that is a common multiple of each of two or more numbers.
Write the 3 multiples that follow the multiples of 2 and 5 shown on the front of the card.
_____
_____

A fraction with a numerator that is greater than or equal to the denominator.
Identify this type of fraction: $2\frac{4}{5}$. Rewrite it as an improper fraction.
_____

MY Vocabulary Cards

Processes & Practices

Lesson 8–9

mixed number

$1\frac{3}{10}$

Lesson 8–3

numerator

$\frac{1}{3}$ $\frac{3}{4}$

Lesson 8–2

prime number

1 × 5   5 × 1   1 × 2   2 × 1

Lesson 8–5

simplest form

$$\frac{8}{12} = \frac{8 \div 4}{12 \div 4} = \frac{2}{3}$$

Ideas for Use

- Create cards to review problem-solving strategies, such as work backward.
- Write the name of each lesson on the front of each card. Write a few study tips for each lesson on the back of each card.

The top number in a fraction. It tells how many of the equal parts are being used.

What does the numerator in the following fraction show? $\frac{9}{20}$

_____

_____

A number that has a whole-number part and a fraction part.

Explain why a mixed number is called "mixed."

_____

_____

_____

A fraction in which the numerator and the denominator have no common factor greater than 1.

Write $\frac{15}{45}$ in its simplest form.

_____

A whole number with exactly two factors— 1 and itself.

Prime comes from the Latin root *primus*, meaning "first." How does this help you understand the definition of *prime number*?

_____

_____

MY Foldable

FOLDABLES® Follow the steps on the back to make your Foldable.

Simplest Form

Three-fourths of voters voted YES!

$\dfrac{3}{4}$

Compare Fractions

Half the population... One-fourth of the population...

$\dfrac{1}{2} > \dfrac{1}{4}$

Mixed Numbers

$1\dfrac{1}{2}$ cups flour

$1\dfrac{1}{2}$

Fractions in Our World

Ways to show fractions

Name _____

Lesson 1
Factors and Multiples

ESSENTIAL QUESTION
How can different fractions name the same amount?

Factors are numbers that are multiplied together to form a product. The two factors that are multiplied together are the **factor pairs** of the product.

Math in My World

Example 1

A chef is arranging 48 strawberries in an array on a tray. Find all the factor pairs of 48 to see what kind of arrays can be made. Draw and label one of the arrays.

Think of multiplication equations that result in a product of 48.

$1 \times 48 = 48$ 1 and 48 are a factor pair of 48.

$2 \times 24 = 48$ 2 and 24 are a factor pair of 48.

$3 \times$ _____ $= 48$ 3 and _____ are a factor pair of 48.

$4 \times$ _____ $= 48$ 4 and _____ are a factor pair of 48.

$6 \times$ _____ $= 48$ 6 and _____ are a factor pair of 48.

48 is a multiple of each of the factors above.

So, the following arrays can be made: _____ × _____, _____ × _____,

_____ × _____, _____ × _____, and _____ × _____.

Lesson 1   485

Example 2

Determine whether 64 is a multiple of 4.

One Way

Divide.

There is no remainder.
So, 4 divides 64 evenly.

So, 64 is a multiple of 4.

Another Way

List multiples of 4.

4, 8, 12, _____, _____,

_____, _____, _____,

_____, _____, _____,

_____, _____, _____,

_____, _____, etc.

You can count by 4 to get to 64.

Guided Practice

Find the factor pairs of each number.

1. 46

   _____ and _____

   _____ and _____

2. 76

   _____ and _____

   _____ and _____

   _____ and _____

For Exercises 3–5, write yes or no.

3. Is 66 a multiple of 6? _____

4. Is 86 a multiple of 4? _____

5. Is 78 a multiple of 3? _____

Would you rather divide or list the multiples for Exercises 3–5? Explain.

486 Chapter 8 Fractions

Name _____

Independent Practice

Find the factor pairs of each number.

6. 66

_____ and _____

_____ and _____

_____ and _____

_____ and _____

7. 54

_____ and _____

_____ and _____

_____ and _____

_____ and _____

8. 59

_____ and _____

9. 58

_____ and _____

_____ and _____

10. 70

_____ and _____

_____ and _____

_____ and _____

_____ and _____

11. 93

_____ and _____

_____ and _____

12. 60 _____

13. 65 _____

14. 37 _____

For Exercises 15–18, write *yes* or *no*.

15. Is 67 a multiple of 5? _____

16. Is 75 a multiple of 3? _____

17. Is 72 a multiple of 4? _____

18. Is 69 a multiple of 6? _____

Circle the number(s) that are multiples of each given number.

19. 3   14   18   27   32   45   60

20. 7   24   38   42   63   71   84

21. 9   30   35   54   82   90   100

Problem Solving

22. There are 49 snacks on a tray. Draw one possible array that can be made.

List all the factor pairs of 49.

_____

23. Processes & Practices 6 **Explain to a Friend** Ella is counting aloud by six. Will she say the number 73? Explain to a friend.

_____

Brain Builders

24. Processes & Practices 7 **Identify Structure** Write two numbers between 55 and 65 that are both multiples of the same number. What factor do they have in common?

_____
_____

Write 2 more numbers less than 55 that have this factor.

_____

25. Processes & Practices 2 **Use Number Sense** Without dividing or listing multiples, how can you tell if 61 is a multiple of 6? Explain two different ways.

_____
_____

26. **Building on the Essential Question** How are factors and multiples related?

_____
_____

488 Chapter 8 Fractions

Name _____

Lesson 1
Factors and Multiples

Homework Helper   Need help? connectED.mcgraw-hill.com

Mr. Carlton has 12 pictures to put on the bulletin board in his classroom. List the factor pairs to find how many different ways Mr. Carlton can arrange the pictures in an array.

Think of multiplication equations that result in a product of 12.

1 × 12 = 12   1 and 12 are a factor pair of 12.
2 × 6 = 12   2 and 6 are a factor pair of 12.
3 × 4 = 12   3 and 4 are a factor pair of 12.

So, Mr. Carlton could arrange the 12 pictures in the following arrays: 1 × 12, 2 × 6, or 3 × 4.

Check
Use models to show possible arrays. The arrays can also be turned the other way.

Practice

Find the factor pairs of each number.

1. 47

_____ and _____

2. 78

_____ and _____
_____ and _____
_____ and _____
_____ and _____

3. 65

_____ and _____
_____ and _____

Lesson 1 My Homework 489

Find the factor pairs of each number.

4. 56

_____ and _____

_____ and _____

_____ and _____

_____ and _____

5. 30

_____ and _____

_____ and _____

_____ and _____

_____ and _____

6. 71

_____ and _____

For Exercises 7–10, write *yes* or *no*.

7. Is 43 a multiple of 7? _____

8. Is 56 a multiple of 6? _____

9. Is 80 a multiple of 4? _____

10. Is 42 a multiple of 3? _____

Brain Builders

11. **Processes & Practices** **Model Math** Janelle is selling lemonade. She has poured 36 cups of lemonade to display. How many different arrays could Janelle use to arrange the lemonade for display? Draw one possible array Janelle could use to display the cups.

_____

_____

Vocabulary Check

12. Write a definition for factor pairs. Then give an example.

_____

_____

13. **Test Practice** Lucas is counting by 8. Which of these numbers will he *not* say?

Ⓐ 32

Ⓑ 56

Ⓒ 72

Ⓓ 84

Name _____

Lesson 2
Prime and Composite Numbers

ESSENTIAL QUESTION
How can different fractions name the same amount?

The factors of a number help us determine whether a number is prime, composite, or neither.

 Math in My World

Example 1

Blair is arranging 18 cupcakes on a shelf. Find the factors of 18. Use arrays to find the factors.

_____ × _____

_____ × _____

_____ × _____

So, the factors of 18 are _____, _____, _____, _____, _____, and _____.

Since 18 has more than two factors, it is a composite number.

Key Concept — Prime and Composite Numbers

Definition	Examples
Prime Number — A whole number with exactly two distinct factors, 1 and itself.	2, 3, 7, 11
Composite Number — A whole number with more than two factors.	4, 6, 10, 20
Neither Prime nor Composite — 1 only has one distinct factor.	1

Online Content at connectED.mcgraw-hill.com

Example 2

Sarah is at the park. She counts 73 kites. Write the factors of 73. Tell whether it is *prime*, *composite*, or *neither*.

The factors of 73 are _____ and _____ .

Since 73 has exactly two distinct factors,

it is a _____ number.

Example 3

Tell whether 1 is *prime*, *composite*, or *neither*.

Find the factors of 1.

1 × 1

Since 1 only has one distinct factor, it is _____ .

Guided Practice

Tell whether each number is *prime*, *composite*, or *neither*.

1. 5

2. 15

3. 21

4. 31

5. 26

6. 61

Identify the smallest prime number. Explain how you know this is the smallest prime number.

492 Chapter 8 Fractions

Name _____

Independent Practice

Tell whether each number is *prime*, *composite*, or *neither*.

7. 1 **8.** 3 **9.** 4 **10.** 14

11. 29 **12.** 41 **13.** 50 **14.** 63

15. 65 **16.** 79 **17.** 84 **18.** 97

19. Circle the prime numbers. Cross out the numbers that are composite or neither.

1	2	3	4	5	6	7	8	9	10
11	12	13	14	15	16	17	18	19	20
21	22	23	24	25	26	27	28	29	30
31	32	33	34	35	36	37	38	39	40
41	42	43	44	45	46	47	48	49	50

20. Write a prime number that is greater than 50.

_____

21. Write a composite number that is greater than 70.

_____

Lesson 2 Prime and Composite Numbers 493

Problem Solving

22. What prime number is greater than 88 and less than 95?

_____

23. Ken is planting vegetables in his garden. He has 20 seeds. Determine whether 20 is a prime or composite number. If it is composite, list all of the ways Ken can arrange the seeds in even rows.

_____

_____

Brain Builders

24. Processes &Practices 4 Model Math Susana is making a quilt by sewing together square pieces of fabric. She has 49 fabric squares. Draw an array to show how she can create a quilt that has the same number of squares in each row and each column. How does the array help you determine whether 49 is prime or composite?

_____

_____

25. Processes &Practices 3 Draw a Conclusion Karrie adds the prime numbers 3 and 7 and gets the composite number 10. She states that the sum of any two prime numbers is a composite number. Is Karrie correct? Explain why or why not.

_____

_____

26. Building on the Essential Question What factors do prime numbers and composite numbers have in common?

_____

_____

Name _____

MY Homework

Lesson 2

Prime and Composite Numbers

Homework Helper

Need help? connectED.mcgraw-hill.com

Patrice is having a tea party. There will be 13 people at the tea party all together. Can Patrice divide the chairs evenly among more than 1 table? Explain.

Find the factors of 13 and decide if 13 is a prime number, a composite number, or neither.

The factors of 13 are 1 and 13. So, 13 is a prime number.

Patrice cannot divide the chairs evenly among more than 1 table because 13 is a prime number.

Type of Number	Definition
prime number	a whole number with exactly two factors, 1 and itself (Examples: 17, 29, 41)
composite number	a whole number with more than two factors (Examples: 8, 30, 56)
neither prime nor composite	a number that has only one distinct factor (Example: 1)

Practice

Tell whether each number is *prime*, *composite*, or *neither*.

1. 16

2. 37

3. 50

_____

_____

_____

4. 41

5. 1

6. 81

_____

_____

_____

Lesson 2 My Homework 495

Tell whether each number is *prime*, *composite*, or *neither*.

7. 0 _____ 8. 11 _____ 9. 90 _____

10. 75 _____ 11. 53 _____ 12. 23 _____

Problem Solving

13. Colby has 16 jars of spices. He wants to arrange them in arrays. What arrays could he use to arrange them?

    _____

14. Winnie has 7 soccer trophies she wants to display in an array. How many different arrays are possible? Explain.

    _____

Brain Builders

15. **Processes & Practices** **Keep Trying** There are 4 prime numbers between 10 and 20; 11, 13, 17, and 19. Are there always the same number of prime numbers between 2 consecutive multiples of 10? Explain.

    _____

    _____

16. Identify the smallest composite number that you can create by multiplying two different composite numbers.

    _____

Vocabulary Check

Draw a line to match the vocabulary term with its example.

17. prime number   • 61

18. composite number   • 21

19. **Test Practice** Which of the following is a prime number?

    Ⓐ 67   Ⓒ 63

    Ⓑ 65   Ⓓ 60

496 **Need more practice?** Download Extra Practice at connectED.mcgraw-hill.com

Check My Progress

Vocabulary Check

1. Circle the **prime** numbers. Draw an X on the **composite** numbers.

   4   7   5   9   11   8   6   14

For Exercises 2–5, write *factor*, *multiple*, or **factor pairs**.

2. 35 is a _____ of 7.

3. 24 and 3 are _____ of 72.

4. 6 is a _____ of 66.

5. 90 is a _____ of 9.

Concept Check

6. Find the factor pairs of 80.

   _____

For Exercises 7–9, write *yes* or *no*.

7. Is 50 a multiple of 5? _____

8. Is 63 a multiple of 4? _____

9. Is 72 a multiple of 3? _____

Tell whether each number is *prime*, *composite*, or *neither*.

10. 4   11. 3   12. 8   13. 1

Problem Solving

14. Mandy is thinking of a prime number that is greater than 41 and less than 47. What is Mandy's number?

_____

15. There are ten desks in a room. Determine whether 10 is prime or composite. If it is composite, list all of the ways the desks can be arranged if they are in even rows.

_____

_____

Brain Builders

16. Fernando has a prime number of books between 21 and 30. He says that if he takes 4 of the books away, he will be able to place the same number of books on each shelf of his bookcase. How many books does Fernando have? How many shelves are on the bookcase? Draw an array to support your answer.

_____

17. If you multiply two prime numbers, will the result be prime or composite? Explain.

_____

_____

18. Test Practice Which sum is prime?

Ⓐ 1 + 3

Ⓑ 2 + 4

Ⓒ 5 + 8

Ⓓ 7 + 9

498 Chapter 8 Fractions

Name _____

Lesson 3 Hands On
Model Equivalent Fractions

ESSENTIAL QUESTION
How can different fractions name the same amount?

The top number on a fraction is the **numerator**. The bottom number on a fraction is the **denominator**. Fractions that represent the same part of a number are **equivalent fractions**.

Build It

Generate two fractions that are equivalent to $\frac{1}{3}$.

1 Model $\frac{1}{3}$.

Place a $\frac{1}{3}$ - tile.

2 Find a fraction equivalent to $\frac{1}{3}$.

Place $\frac{1}{6}$ - tiles below the $\frac{1}{3}$ - tile to equal the length of the $\frac{1}{3}$ - tile.

How many $\frac{1}{6}$ - tiles did you place? _____

So, $\frac{1}{3}$ and $\frac{2}{6}$ are equivalent fractions.

3 Find another fraction equivalent to $\frac{1}{3}$.

Place $\frac{1}{12}$ - tiles below the $\frac{1}{6}$ - tiles to equal the length of the $\frac{1}{3}$ - tile.

How many $\frac{1}{12}$ - tiles did you place? _____

So, $\frac{1}{3}$ and $\frac{4}{12}$ are equivalent fractions.

So, $\frac{1}{3}$, $\frac{\square}{6}$, and $\frac{\square}{12}$ are equivalent fractions.

Online Content at connectED.mcgraw-hill.com

Try It

Generate two fractions that are equivalent to $\frac{1}{4}$.

1. The first number line is divided into fourths. Plot $\frac{1}{4}$ on the number line.

2. The second number line is divided into eighths.

What fraction is at the same location as $\frac{1}{4}$? $\frac{\square}{\square}$

Plot this fraction on the number line.

3. The third number line is divided into twelfths.

What fraction is at the same location as $\frac{1}{4}$? $\frac{\square}{\square}$

Plot this fraction on the number line.

So, two fractions that are equivalent to $\frac{1}{4}$ are $\frac{\square}{\square}$ and $\frac{\square}{\square}$.

Talk About It

1. Processes & Practices 8 Look for a Pattern The table shows some equivalent fractions. Study the table. Describe the pattern between the numerators and denominators of two equivalent fractions.

Equivalent Fractions	
$\frac{1}{3}$	$\frac{2}{6}$
$\frac{1}{3}$	$\frac{4}{12}$
$\frac{1}{4}$	$\frac{2}{8}$
$\frac{1}{4}$	$\frac{3}{12}$

_____

_____

_____

2. Processes & Practices 3 Draw a Conclusion Without using fraction tiles or number lines, determine whether $\frac{1}{2}$ and $\frac{3}{6}$ are equivalent fractions. Explain.

_____

500 Chapter 8 Fractions

Name ....

Practice It

Recognize whether the fractions are equivalent. Write *yes* or *no*. Use fraction tiles or number lines.

3. $\frac{2}{4}$ and $\frac{6}{12}$ 	4. $\frac{6}{8}$ and $\frac{5}{10}$ 	5. $\frac{2}{3}$ and $\frac{3}{5}$

6. $\frac{9}{12}$ and $\frac{3}{4}$ 	7. $\frac{4}{6}$ and $\frac{8}{12}$ 	8. $\frac{2}{3}$ and $\frac{6}{10}$

Generate two equivalent fractions for each fraction. Use fraction tiles or number lines.

9. $\frac{2}{4}$ 	10. $\frac{2}{6}$ 	11. $\frac{4}{8}$

12. $\frac{5}{10}$ 	13. $\frac{1}{3}$ 	14. $\frac{2}{3}$

Lesson 3 Hands On: Model Equivalent Fractions 501

Apply It

15. Processes & Practices 4 **Model Math** There were 10 baked goods in a basket. Four of them were sold. Write a fraction to show the part of the baked goods that were not sold. Then write a fraction that is equivalent to this number.

_____

16. Two-thirds of a jar of peanut butter has been used. Write an equivalent fraction.

_____

17. A jar has marbles in it. Three-tenths of the marbles are red. Five-tenths of the marbles are blue. Two-tenths of the marbles are green. Which of these fractions is equivalent to four-eighths?

_____

18. Processes & Practices 2 **Use Number Sense** Daria used fraction tiles to show that $\frac{3}{5}$ is equivalent to $\frac{6}{10}$. Compare the number and size of fraction tiles needed to model each fraction.

_____

_____

19. Processes & Practices 1 **Make Sense of Problems**

Complete the equation.

Write About It

20. Write a real-world example of equivalent fractions.

_____

502 Chapter 8 Fractions

Name _____

MY Homework

Lesson 3

Hands On: Model Equivalent Fractions

Homework Helper

Need help? connectED.mcgraw-hill.com

Determine whether $\frac{1}{2}$ is equivalent to $\frac{3}{6}$.

One Way Use fraction tiles.

1 Model $\frac{1}{2}$.

2 Model $\frac{3}{6}$.

Line up three $\frac{1}{6}$-fraction tiles below the $\frac{1}{2}$-fraction tile.

It takes three $\frac{1}{6}$-tiles, so the fraction is $\frac{3}{6}$.

Since they are the same length, the fractions are equivalent.

So, $\frac{1}{2} = \frac{3}{6}$.

Another Way Use number lines.

1 Divide the first number line into halves.

2 Divide the second number line into sixths.

3 Count the number of sixths that are in one half.

The number lines show that $\frac{1}{2}$ and $\frac{3}{6}$ are at the same point.

So, they are equivalent fractions.

Lesson 3 My Homework 503

Practice

Recognize whether the fractions are equivalent. Write *yes* or *no*. Use fraction tiles or number lines.

1. $\frac{3}{5}$ and $\frac{6}{8}$

2. $\frac{4}{5}$ and $\frac{5}{6}$

3. $\frac{2}{4}$ and $\frac{6}{12}$

4. $\frac{2}{3}$ and $\frac{4}{6}$

5. $\frac{8}{12}$ and $\frac{4}{6}$

6. $\frac{5}{6}$ and $\frac{9}{10}$

Generate two equivalent fractions for each fraction. Use fraction tiles or number lines.

7. $\frac{1}{3}$

8. $\frac{8}{12}$

9. $\frac{3}{4}$

Problem Solving

10. **Processes &Practices 3** **Justify Conclusions** Francie lives $\frac{1}{5}$ mile from the school. Jake lives $\frac{2}{10}$ mile from the school. Do they live the same distance from the school? Explain.

_____

_____

Vocabulary Check

Draw a line to match the vocabulary term with its example.

11. numerator • $\frac{6}{10}$ and $\frac{3}{5}$

12. denominator • the number 1 in $\frac{1}{4}$

13. equivalent fractions • the number 4 in $\frac{1}{4}$

Chapter 8 Fractions

Name _____

Lesson 4
Equivalent Fractions

ESSENTIAL QUESTION
How can different fractions name the same amount?

You have used models and number lines to find equivalent fractions. Multiplication can also be used to find equivalent fractions. Fractions that have the same numerator and denominator are equivalent to one whole.

Math in My World

Example 1

A recipe for spaghetti and meatballs calls for $\frac{3}{4}$ pound of ground beef. Find two fractions that are equivalent to $\frac{3}{4}$.

One Way Use models.

Model $\frac{3}{4}$.

Place six $\frac{1}{8}$-tiles below the $\frac{1}{4}$-tiles. Place nine $\frac{1}{12}$-tiles below the $\frac{1}{8}$-tiles.

The $\frac{1}{8}$-tiles are smaller than the $\frac{1}{4}$-tiles, so there are more of them.

The $\frac{1}{12}$-tiles are smaller than the $\frac{1}{4}$-tiles and $\frac{1}{8}$-tiles, so there are more of them.

Another Way Use multiplication.

Multiply the numerator and denominator by the same number.

$\frac{3}{4} = \frac{3 \times 2}{4 \times 2} = \frac{6}{8}$ $\frac{3}{4} = \frac{3 \times 3}{4 \times 3} = \frac{9}{12}$

So, two fractions that are equivalent to $\frac{3}{4}$ are $\frac{6}{8}$ and $\frac{9}{12}$.

Online Content at connectED.mcgraw-hill.com

You can use multiplication to write a fraction with a denominator of 10 as an equivalent fraction with a denominator of 100.

Example 2

Liliana has a container of dry beans. The table shows the fraction of each type of bean. Which type of bean shows a fraction that is equivalent to $\frac{3}{10}$ of the container?

Bean	Fraction
Black	$\frac{20}{100}$
Kidney	$\frac{10}{100}$
Garbanzo	$\frac{40}{100}$
Lima	$\frac{30}{100}$

Each fraction in the table has a denominator of 100. Use multiplication to find which fraction is equivalent to $\frac{3}{10}$.

Multiply.

$$\frac{3}{10} = \frac{3 \times \boxed{10}}{10 \times \boxed{10}} = \frac{\boxed{}}{\boxed{}}$$

So, $\frac{3}{10}$ of the container has _____ beans.

Guided Practice

Write the fraction for the part that is shaded. Then find an equivalent fraction.

1.

   _____

2.

   _____

Talk MATH

Tell why $\frac{3}{4}$, $\frac{6}{8}$, and $\frac{9}{12}$ are equivalent fractions. Give an example of another set of three equivalent fractions.

Name ....

Independent Practice

Write the fraction for the part that is shaded. Then find an equivalent fraction.

3.

4.

_____

_____

5.

6.

_____

_____

Circle the two fractions that are equivalent in each set of fractions.

7. $\frac{1}{4}$ $\frac{3}{12}$ $\frac{2}{6}$

8. $\frac{2}{10}$ $\frac{2}{100}$ $\frac{20}{100}$

9. $\frac{3}{5}$ $\frac{6}{8}$ $\frac{6}{10}$

10. $\frac{1}{100}$ $\frac{10}{100}$ $\frac{1}{10}$

11. $\frac{2}{3}$ $\frac{5}{6}$ $\frac{8}{12}$

12. $\frac{6}{10}$ $\frac{40}{100}$ $\frac{4}{10}$

Algebra Find each unknown.

13. $\frac{1}{4} = \frac{\square}{8}$

14. $\frac{7}{10} = \frac{\square}{100}$

15. $\frac{3}{5} = \frac{6}{\square}$

Lesson 4 Equivalent Fractions 507

Problem Solving

Female deer, called does, usually have 1 to 3 young at a time. Young deer are called fawns. The male deer, called bucks, grow antlers in the spring.

16. Suppose 10 deer are traveling together. If 4 of them are does and 3 are fawns, what fraction represents the number of bucks? Write another equivalent fraction that represents this number.

_____

17. A roller coaster has 10 cars. Seven of the cars are green. Write two equivalent fractions for the part of the cars that is green.

_____

Brain Builders

18. **Processes & Practices 4** **Model Math** Javier has 4 juice boxes. Three are grape flavored. Write two equivalent fractions that describe the part of the juice boxes that is grape. Draw a model for each fraction to show that they are equivalent.

19. **Processes & Practices 2** **Use Number Sense** Identify two fractions that are equivalent to $\frac{25}{100}$.

_____

20. **Building on the Essential Question** Describe two methods you can use to check to see if two fractions are equivalent.

_____
_____
_____

508 Chapter 8 Fractions

Name

MY Homework

Lesson 4

Equivalent Fractions

Homework Helper

Need help? connectED.mcgraw-hill.com

Write the fraction for the part that is shaded. Then find two equivalent fractions.

1 Find the fraction that represents the shaded part.

$\dfrac{4}{8}$ ← number of shaded parts
← total number of parts

2 Find equivalent fractions.

Multiply the numerator and denominator by the same number, for example, 2.

$$\dfrac{4 \times 2}{8 \times 2} = \dfrac{8}{16}$$

Multiply the numerator and denominator by another number, for example, 3.

$$\dfrac{4 \times 3}{8 \times 3} = \dfrac{12}{24}$$

So, the fraction represented by the circle is $\dfrac{4}{8}$.

Two equivalent fractions are $\dfrac{8}{16}$ and $\dfrac{12}{24}$.

Practice

Write the fraction for the part that is shaded. Then find an equivalent fraction.

1. _____

2. _____

Lesson 4 My Homework 509

Write the fraction for the part that is shaded. Then find an equivalent fraction.

3. _____

_____

4. _____

_____

Find an equivalent fraction for each fraction.

5. $\frac{20}{100}$ _____

6. $\frac{2}{8}$ _____

7. $\frac{90}{100}$ _____

Algebra Find each unknown.

8. $\frac{6}{12} = \frac{x}{2}$

9. $\frac{3}{10} = \frac{x}{100}$

10. $\frac{5}{8} = \frac{10}{x}$

x = _____

x = _____

x = _____

Brain Builders

11. **Processes &Practices 1** **Use Number Sense** Janie has 12 pieces of fruit. Eight of the pieces of fruit are bananas. Write three fractions that describe the fraction of fruit that is bananas.

_____

12. A box contains 4 red pencils and 6 black pencils. What fraction of the pencils are red? What fraction of pencils are black? Draw a model that shows the fraction of each color of pencil.

_____

13. **Test Practice** Laura delivers newspapers. She spent $\frac{4}{12}$ of her savings on a new CD. Which equivalent fraction shows the amount Laura spent?

Ⓐ $\frac{1}{9}$

Ⓒ $\frac{2}{8}$

Ⓑ $\frac{1}{3}$

Ⓓ $\frac{2}{3}$

Name _____

Lesson 5
Simplest Form

ESSENTIAL QUESTION
How can different fractions name the same amount?

A fraction is in **simplest form** when its numerator and denominator have no common factor other than 1. The simplest form of a fraction is equivalent to the fraction.

Math in My World

Example 1

There are 12 ingredients in a bread recipe. Esteban needs to buy 8 out of 12, or $\frac{8}{12}$, of the ingredients. Write $\frac{8}{12}$ in simplest form.

To write a fraction in simplest form, divide both the numerator and the denominator by the greatest common factor. The **greatest common factor** is the greatest of the common factors of two or more numbers.

 Find the common factors of 8 and 12.

Factors of 8: 1, 2, 4, 8   Factors of 12: 1, 2, 3, 4, 6, 12

Common factors: ____, ____, ____   Circle the greatest common factor.

 Divide by the greatest common factor.

$\dfrac{8 \div 4}{12 \div 4} = \dfrac{\Box}{\Box}$ The numbers 2 and 3 have no common factor other than 1.

So, $\frac{8}{12}$ in simplest form is $\dfrac{\Box}{\Box}$.

Online Content at connectED.mcgraw-hill.com

To write a fraction in simplest form, you can also divide by common factors.

Example 2

A newborn Koala is $\frac{6}{8}$ inch long. Write the fraction in simplest form.

1 Write the fraction.

$\frac{\square}{\square}$

2 Divide by common factors.

$\frac{6}{8} = \frac{6 \div 2}{8 \div 2} = \frac{\square}{\square}$ A common factor of 6 and 8 is 2.

Helpful Hint
Divide until the numerator and the denominator have no common factor other than 1.

So, $\frac{6}{8}$ simplifies to $\frac{\square}{\square}$.   A newborn Koala is $\frac{\square}{4}$ inch long.

Guided Practice

Write each fraction in simplest form. If it is in simplest form, write *simplest form*.

1. $\frac{2}{4}$

2. $\frac{6}{10}$

Talk MATH
How do you know that $\frac{4}{5}$ is in simplest form?

3. $\frac{5}{6}$

4. $\frac{4}{12}$

512 Chapter 8 Fractions

Name ....

Independent Practice

Write each fraction in simplest form. If it is in simplest form, write *simplest form*.

5. $\dfrac{2}{10}$

6. $\dfrac{3}{12}$

7. $\dfrac{7}{10}$

8. $\dfrac{6}{12}$

9. $\dfrac{8}{12}$

10. $\dfrac{2}{3}$

11. $\dfrac{8}{10}$

12. $\dfrac{5}{10}$

13. $\dfrac{3}{4}$

14. $\dfrac{4}{5}$

15. $\dfrac{1}{2}$

16. $\dfrac{4}{8}$

17. $\dfrac{20}{100}$

18. $\dfrac{40}{100}$

19. $\dfrac{50}{100}$

Algebra Find each unknown.

20. $\dfrac{4}{10} = \dfrac{\blacksquare}{5}$

21. $\dfrac{6}{8} = \dfrac{3}{\blacksquare}$

22. $\dfrac{60}{100} = \dfrac{\blacksquare}{5}$

$\blacksquare =$

$\blacksquare =$

$\blacksquare =$

Lesson 5 Simplest Form 513

Problem Solving

The table lists the lengths of some objects in a classroom. What part of a foot is each object? Write as a fraction in simplest form.

23. pencil _____   24. crayon _____

25. scissors _____   26. pencil box _____

For Exercise 27, write the answer as a fraction in simplest form.

27. Ricardo has made 4 out of the 12 recipes in his cookbook. What fraction of the recipes has he made?

_____

Brain Builders

28. **Processes &Practices 1** **Make Sense of Problems** A bag contains 6 yellow marbles, 1 white marble, and 3 red marbles. What fraction of the marbles is yellow? Write the answer in simplest form.

_____

29. **Processes &Practices 7** **Identify Structure** Write three fractions that are in simplest form. Explain how you know.

_____
_____

30. **Building on the Essential Question** How can you use the simplest form of fractions to determine if two fractions are equivalent? Explain.

_____
_____

514 Chapter 8 Fractions

Name ....

MY Homework

Lesson 5

Simplest Form

Homework Helper

Need help? connectED.mcgraw-hill.com

Sophie found 12 golf balls. Six of them are yellow. What fraction of the golf balls are yellow? Write the fraction in its simplest form.

1 Write the fraction.

6 out of 12 golf balls are yellow.

The fraction is $\frac{6}{12}$.

2 Find the common factors of 6 and 12.

Circle the greatest common factor.

Factors of 6: 1, 2, 3, ⑥

Factors of 12: 1, 2, 3, 4, ⑥, 12

Helpful Hint

A fraction is in its simplest form when its numerator and denominator have no common factor other than 1.

3 Divide by the greatest common factor.

$$\frac{6 \div 6}{12 \div 6} = \frac{1}{2}$$

So, $\frac{1}{2}$ of the golf balls are yellow.

Practice

Write each fraction in simplest form. If it is in simplest form, write *simplest form*.

1. $\frac{3}{12}$

2. $\frac{6}{8}$

3. $\frac{7}{12}$

4. $\frac{4}{10}$

5. $\frac{20}{100}$

6. $\frac{2}{6}$

Write each fraction in simplest form. If it is in simplest form, write *simplest form*.

7. $\dfrac{5}{10}$ _____

8. $\dfrac{4}{12}$ _____

9. $\dfrac{2}{8}$ _____

10. $\dfrac{3}{100}$ _____

11. $\dfrac{70}{100}$ _____

12. $\dfrac{2}{5}$ _____

Brain Builders

13. **Processes &Practices** **Make Sense of Problems** Latitia had 12 marbles. After she gave 2 marbles to Emilia, she has $\dfrac{10}{12}$ of her marbles left. What fraction of the marbles did Latitia give away? Write in simplest form.

_____

14. Ryan has 8 kittens. Two are white. Julie has 4 kittens. The fraction of kittens that are *not* white is the same for Ryan and Julie. What fraction of Julie's kittens are *not* white? Write in simplest form.

_____

Vocabulary Check

Write a vocabulary term to complete each sentence.

greatest common factor   simplest form

15. When the only common factor of the numerator and denominator is 1, a fraction is in _____.

16. To simplify a fraction, divide the numerator and denominator by their _____.

17. **Test Practice** What is the unknown denominator if $\dfrac{9}{12} = \dfrac{3}{\blacksquare}$?

Ⓐ 8

Ⓑ 6

Ⓒ 4

Ⓓ 3

Name ....

Lesson 6
Compare and Order Fractions

ESSENTIAL QUESTION
How can different fractions name the same amount?

To compare fractions, create equivalent fractions with the same denominators or the same numerators. Use the **least common multiple**, or the least multiple common to sets of multiples.

Math in My World

Example 1

Ramon has an insect collection. The table shows the lengths of four insects in his collection. Which is longer, a mosquito or a whirligig beetle?

Insect	Length (in.)
Mosquito	$\frac{1}{4}$
Field cricket	$\frac{5}{8}$
Whirligig beetle	$\frac{3}{8}$
Lightning bug	$\frac{1}{2}$

 Find the least common multiple of the denominators.

Multiples of 4: 4, **8**, 12, 16, ... ← mosquito

Multiples of 8: **8**, 16, 24, ... ← whirligig beetle

The least common multiple of the denominators is _____.

 Create equivalent fractions that use 8 as the denominator.

Mosquito

$$\frac{1}{4} = \frac{1 \times 2}{4 \times 2} = \frac{2}{8}$$

Whirligig Beetle

$$\frac{3}{8} = \frac{3 \times 1}{8 \times 1} = \frac{3}{8}$$

 Compare the numerators.
When the denominators are the same, the fraction with the greater numerator is the greater fraction.

$\frac{2}{8} < \frac{3}{8}$ ← Of the numbers 2 and 3, 3 is closer to 8. So, 2 out of 8 is smaller than 3 out of 8.

So, a _____ is longer.

Online Content at connectED.mcgraw-hill.com

Example 2

The table shows the cooking times needed for different foods. Order the foods from *least* to *greatest* cooking times.

Cooking Times	
Food	Time
Rice	$\frac{1}{4}$ hour
Lasagna	$\frac{5}{6}$ hour
Enchiladas	$\frac{2}{3}$ hour

 Find the least common multiple of the numerators.

Multiples of 1: 1, 2, 3, 4, 5, 6, 7, 8, 9, **10**, 11, ... ← rice

Multiples of 5: 5, **10**, 15, ... ← lasagna

Multiples of 2: 2, 4, 6, 8, **10**, 12, ... ← enchiladas

The least common multiple of the numerators is _____.

 Generate equivalent fractions that use 10 as the numerator.

Rice: $\frac{1}{4} = \frac{1 \times 10}{4 \times 10} = \frac{10}{40}$

Lasagna: $\frac{5}{6} = \frac{5 \times 2}{6 \times 2} = \frac{10}{12}$

Enchiladas: $\frac{2}{3} = \frac{2 \times 5}{3 \times 5} = \frac{10}{15}$

Compare the denominators.

When the numerators are the same, the fraction with the greatest denominator is the least fraction.

$\frac{10}{40} < \frac{10}{15} < \frac{10}{12}$ ← 10 is much closer to 12 than it is to 40, so 10 out of 40 is smaller than 10 out of 12.

So, from *least* to *greatest* cooking times, the foods are _____, _____, and _____.

Check

The models show $\frac{1}{4} < \frac{2}{3} < \frac{5}{6}$.

$\frac{1}{4}$ $\frac{2}{3}$ $\frac{5}{6}$

Talk MATH

Explain how to compare $\frac{7}{12}$ and $\frac{2}{6}$.

Guided Practice

Compare. Use >, <, or =.

1. $\frac{3}{4} \bigcirc \frac{1}{2}$

2. $\frac{3}{6} \bigcirc \frac{3}{4}$

518 Chapter 8 Fractions

Name

Independent Practice

Compare. Use >, <, or =. Check your answer using fraction tiles or number lines.

3. $\frac{2}{6} \bigcirc \frac{1}{3}$

4. $\frac{3}{5} \bigcirc \frac{5}{6}$

5. $\frac{4}{5} \bigcirc \frac{8}{10}$

6. $\frac{2}{3} \bigcirc \frac{1}{3}$

7. $\frac{4}{10} \bigcirc \frac{1}{2}$

8. $\frac{5}{8} \bigcirc \frac{2}{3}$

9. $\frac{3}{4} \bigcirc \frac{1}{3}$

10. $\frac{2}{3} \bigcirc \frac{6}{9}$

11. $\frac{1}{2} \bigcirc \frac{3}{4}$

Order from *least* to *greatest*. Check your answer using fraction tiles or number lines.

12. $\frac{4}{6}, \frac{1}{3}, \frac{3}{3}$

13. $\frac{3}{4}, \frac{2}{3}, \frac{7}{8}$

14. $\frac{3}{10}, \frac{3}{5}, \frac{3}{4}$

15. $\frac{1}{6}, \frac{2}{5}, \frac{3}{4}$

16. $\frac{3}{8}, \frac{2}{5}, \frac{2}{10}$

17. $\frac{3}{5}, \frac{2}{4}, \frac{1}{3}$

Lesson 6 Compare and Order Fractions

Problem Solving

18. Which meat makes up the most of Mr. Collin's sandwich?

_____

19. Processes & Practices 1 **Plan Your Solution** Aisha ate $\frac{1}{4}$ of the carrots in the bag. Enrique ate $\frac{3}{12}$ of the carrots in the bag. Compare the amounts of carrots they each ate.

_____

Brain Builders

20. Processes & Practices 2 **Find the Error** Makayla said that $\frac{3}{4}$ of Figure 1 > $\frac{2}{4}$ of Figure 2. Find and correct her mistake.

Figure 1   Figure 2

_____

_____

21. Processes & Practices 2 **Reason** Write three fractions that are *not* greater than $\frac{1}{2}$. Explain how you determined your fractions.

_____

_____

22. Building on the Essential Question How can I compare two fractions with the same numerator? Draw a model to support your reasoning.

_____

_____

520 Chapter 8 Fractions

MY Homework

Lesson 6

Compare and Order Fractions

Homework Helper

Need help? connectED.mcgraw-hill.com

Ellen has three cans of paint that are the same size. The can of blue paint is $\frac{2}{3}$ full. The can of green paint is $\frac{3}{4}$ full, and the can of yellow paint is $\frac{1}{2}$ full. Order the paint colors from least to greatest amounts.

Compare $\frac{2}{3}$, $\frac{3}{4}$, and $\frac{1}{2}$.

1 Find the least common multiple of the denominators.

Circle the least common multiple.

Multiples of 2: 2, 4, 6, 8, 10, ⓬, 14

Multiples of 3: 3, 6, 9, ⓬, 15

Multiples of 4: 4, 8, ⓬, 16

The least common multiple is 12.

2 Create equivalent fractions.

Multiply to find equivalent fractions with 12 as the denominator.

$\frac{2 \times 4}{3 \times 4} = \frac{8}{12}$ $\frac{3 \times 3}{4 \times 3} = \frac{9}{12}$ $\frac{1 \times 6}{2 \times 6} = \frac{6}{12}$

Helpful Hint

You can also use models to compare the fractions.

$\frac{1}{2} < \frac{2}{3} < \frac{3}{4}$

3 Compare and order.

$\frac{6}{12} < \frac{8}{12} < \frac{9}{12}$

So, $\frac{1}{2} < \frac{2}{3} < \frac{3}{4}$.

From *least* to *greatest*, the amount of each color of paint is yellow, blue, and green.

Practice

Compare. Use >, <, or =.

1. $\frac{1}{2}$ ◯ $\frac{1}{3}$
2. $\frac{5}{12}$ ◯ $\frac{1}{4}$
3. $\frac{4}{5}$ ◯ $\frac{8}{10}$
4. $\frac{7}{10}$ ◯ $\frac{4}{5}$
5. $\frac{1}{5}$ ◯ $\frac{2}{10}$
6. $\frac{2}{5}$ ◯ $\frac{2}{8}$
7. $\frac{9}{10}$ ◯ $\frac{7}{8}$
8. $\frac{3}{10}$ ◯ $\frac{4}{8}$
9. $\frac{1}{4}$ ◯ $\frac{6}{12}$

Order from *least* to *greatest*.

10. $\frac{4}{8}, \frac{1}{3}, \frac{2}{3}$ _____
11. $\frac{5}{6}, \frac{7}{12}, \frac{3}{4}$ _____
12. $\frac{1}{2}, \frac{7}{8}, \frac{2}{8}$ _____
13. $\frac{1}{3}, \frac{1}{4}, \frac{5}{6}$ _____

Brain Builders

14. Patti has two glue sticks that are partially used. One has $\frac{1}{5}$ left, and one has $\frac{3}{8}$ left. Which glue stick has more glue? Explain.

_____

15. **Processes & Practices 6** **Be Precise** Lola measures three buttons for a shirt she is making. One button is $\frac{1}{8}$ inch, one is $\frac{3}{8}$ inch, and one is $\frac{1}{4}$ inch. The button hole on the shirt is $\frac{2}{6}$ inch. Which of the buttons will fit through the button hole? Explain.

_____

Vocabulary Check

16. What is the least common multiple of 3 and 8? _____

17. **Test Practice** Which fraction is *not* greater than $\frac{1}{2}$?

Ⓐ $\frac{7}{8}$ Ⓑ $\frac{4}{6}$ Ⓒ $\frac{3}{5}$ Ⓓ $\frac{2}{5}$

Name ....

Lesson 7
Use Benchmark Fractions to Compare and Order

ESSENTIAL QUESTION
How can different fractions name the same amount?

Benchmark fractions are common fractions, such as $\frac{1}{2}$, that are often used to compare and order fractions.

Math in My World

Example 1

A recipe for granola uses $\frac{5}{8}$ cup of raisins and $\frac{1}{6}$ cup of dried cranberries. Is there a greater amount of raisins or cranberries?

Compare $\frac{5}{8}$ and $\frac{1}{6}$.

 Use $\frac{1}{2}$ as a benchmark fraction.

You know that $\frac{1}{2}$ is equivalent to $\frac{4}{8}$ and $\frac{3}{6}$.

 Compare. $\frac{5}{8} > \frac{1}{2}$ and $\frac{1}{6} < \frac{1}{2}$.

So, $\frac{5}{8}$ ◯ $\frac{1}{6}$.

There is a greater amount of _____ in the granola recipe.

Online Content at connectED.mcgraw-hill.com

Lesson 7 523

Example 2

List $\frac{7}{8}$, $\frac{1}{2}$, and $\frac{1}{3}$ in order from *least* to *greatest*.

 Use $\frac{1}{2}$ as a benchmark fraction.

You know that $\frac{1}{2}$ is equivalent to $\frac{4}{8}$ and $\frac{3}{6}$.

$\frac{1}{3}$ is equivalent to $\frac{2}{6}$.

 Compare. $\frac{7}{8} > \frac{1}{2}$ and $\frac{1}{3} < \frac{1}{2}$.

So, from *least* to *greatest*, the fractions are $\frac{1}{3}, \frac{1}{2}, \frac{7}{8}$.

$\frac{\square}{\square} < \frac{\square}{\square} < \frac{\square}{\square}$

Talk MATH

Explain how you know $\frac{1}{8} < \frac{1}{2}$.

Guided Practice

Use a benchmark fraction to compare.
Write >, <, or =.

1. $\frac{3}{4}$ ◯ $\frac{1}{2}$

2. $\frac{3}{6}$ ◯ $\frac{3}{4}$

524 Chapter 8 Fractions

Name _____

Lesson 7

Use Benchmark Fractions to Compare and Order

Homework Helper

Need help? connectED.mcgraw-hill.com

Connor has completed $\frac{3}{8}$ of his homework. Aidan has finished $\frac{5}{6}$ of his homework. Who is closer to being finished with his homework?

Use benchmark fractions, or common fractions, to compare $\frac{3}{8}$ and $\frac{5}{6}$.

 Use $\frac{1}{2}$ as a benchmark fraction.

You know that $\frac{1}{2}$ is equivalent to $\frac{4}{8}$ and $\frac{3}{6}$.

2 Compare.

$\frac{3}{8}$ is less than $\frac{1}{2}$, and $\frac{5}{6}$ is greater than $\frac{1}{2}$.

So, Aidan is closer to finishing his homework than Connor.

Practice

Use a benchmark fraction to compare. Use >, <, or =.

1. $\frac{3}{8}$ ◯ $\frac{3}{4}$

2. $\frac{4}{6}$ ◯ $\frac{3}{8}$

3. $\frac{1}{8}$ ◯ $\frac{5}{6}$

Lesson 7 My Homework 527

Use a benchmark fraction to compare. Use >, <, or =.

4. $\frac{1}{3}$ ◯ $\frac{3}{4}$

5. $\frac{3}{5}$ ◯ $\frac{2}{6}$

6. $\frac{3}{6}$ ◯ $\frac{2}{5}$

7. $\frac{2}{4}$ ◯ $\frac{2}{3}$

Order from *least* to *greatest*.

8. $\frac{1}{2}, \frac{1}{4}, \frac{2}{3}$ _____

9. $\frac{3}{6}, \frac{3}{4}, \frac{2}{8}$ _____

Brain Builders

10. Bailey has an hour to get ready for school. She spends $\frac{3}{6}$ of that time getting dressed. She spends $\frac{1}{4}$ of that time eating breakfast. She spends $\frac{1}{5}$ of that time packing her backpack. Which activity does Bailey spend the greatest amount of time doing? Explain.

_____

11. **Processes & Practices** **2** **Use Number Sense** Zane eats $\frac{3}{8}$ of a pizza. Later he eats $\frac{1}{2}$ of the same pizza. Did he eat more the first or second time? Explain.

_____

Vocabulary Check

12. How do I use benchmark fractions?

_____

13. **Test Practice** Which fraction is not greater than $\frac{2}{5}$?

Ⓐ $\frac{3}{10}$ Ⓑ $\frac{7}{10}$ Ⓒ $\frac{2}{4}$ Ⓓ $\frac{3}{6}$

Need more practice? Download Extra Practice at connectED.mcgraw-hill.com

Check My Progress

Vocabulary Check

1. Write two **equivalent fractions**. _____

2. Write the fraction that has a 4 in the **numerator** and a 6 in the **denominator**.

   _____

3. Write the fraction from Exercise 2 in **simplest form**. _____

4. Explain the difference between a **greatest common factor** and a **least common multiple**.

   _____
   _____
   _____

Concept Check

Recognize whether the fractions are equivalent. Write *yes* or *no*. Use fraction tiles or number lines.

5. $\frac{1}{4}$ and $\frac{3}{12}$

6. $\frac{4}{6}$ and $\frac{6}{12}$

7. $\frac{5}{10}$ and $\frac{3}{6}$

Generate two equivalent fractions for each fraction. Use fraction tiles or number lines.

8. $\frac{2}{8}$

9. $\frac{1}{3}$

10. $\frac{2}{4}$

Problem Solving

11. Adina read 60 out of 100 pages in a comic book. Write the fraction of the pages she read in simplest form.

_____

Brain Builders

12. The table shows how much time each student needs to finish an art project. Does Simón need more or less time than Phil? Explain.

Student Time	
Simón	$\frac{1}{3}$ hour
Phil	$\frac{3}{4}$ hour

_____

Janice needs less time than either Simón or Phil. How much time could Janice need to complete her art project?

_____

13. Write the fraction for the part that is shaded. Then find an equivalent fraction.

_____

Draw a model of the equivalent fraction. Explain why the fractions are equivalent.

_____

14. **Test Practice** Which fraction is in simplest form?

Ⓐ $\frac{2}{10}$ Ⓒ $\frac{4}{12}$

Ⓑ $\frac{3}{12}$ Ⓓ $\frac{3}{8}$

Name _____

Lesson 8
Problem-Solving Investigation
STRATEGY: Use Logical Reasoning

ESSENTIAL QUESTION
How can different fractions name the same amount?

Learn the Strategy

Marla used flour, sugar, and brown sugar to make a treat. Use the clues below to find the amount of each ingredient she used.
- The amounts were $\frac{3}{4}$ cup, $\frac{1}{4}$ cup, and $\frac{2}{3}$ cup.
- She used more flour than sugar.
- She used more sugar than brown sugar.

1 Understand

What facts do you know?

Marla used more _____ than sugar and more sugar than _____.

What do you need to find?

the amount of each _____ she used

2 Plan

I can use logical reasoning to solve the problem.

3 Solve

The order of the ingredients from greatest to least is _____, _____, and _____.

The greatest fraction is $\frac{3}{4}$. So, Marla used $\frac{3}{4}$ cup of flour. $\frac{2}{3} > \frac{1}{4}$
So, she used $\frac{2}{3}$ cup of sugar and $\frac{1}{4}$ cup of brown sugar.

4 Check

Does your answer make sense? Explain.

_____

Online Content at connectED.mcgraw-hill.com

Lesson 8 531

Practice the Strategy

Ameer is making salsa using tomatoes, onions, and black beans. Use the clues below to find the amount of each ingredient he used.

- The amounts are: $\frac{1}{2}$ cup, $\frac{3}{4}$ cup, and $\frac{1}{3}$ cup.
- He is using more tomatoes than black beans and more black beans than onions.

1 Understand

What facts do you know?

_____

_____

What do you need to find?

_____

2 Plan

_____

_____

3 Solve

4 Check

Does your answer make sense? Explain.

_____

Name _____

Lesson 9
Mixed Numbers

ESSENTIAL QUESTION
How can different fractions name the same amount?

A **mixed number** has a whole-number part and a fraction part. It represents an amount greater than one whole.

There is more than 1 sandwich.

There are $1\frac{1}{2}$ sandwiches.

The line is more than 2 inches long.

The line is $2\frac{1}{4}$ inches long.

$1 + \frac{1}{2} = 1\frac{1}{2}$

$1 + 1 + \frac{1}{4} = 2\frac{1}{4}$

Math in My World

Example 1

Caden had 3 apples. He cut off half of an apple to eat. What mixed number represents the amount of apples left?

Count the wholes. Then count the parts.

There are 2 whole apples and $\frac{1}{2}$ of an apple left.

$1 + 1 + \frac{1}{2} = 2\frac{1}{2}$

So, $\boxed{}\frac{\boxed{}}{\boxed{}}$ of the apples are left.

Online Content at connectED.mcgraw-hill.com

You can decompose a mixed number into a sum of whole numbers and unit fractions. Recall that a unit fraction is a fraction with a numerator of 1.

Example 2

Write the length of this leaf as a mixed number. Write an equation to show the mixed number.

 Count the wholes. The leaf is 3 whole inches long.

Count the parts. The leaf is $\frac{3}{4}$ of an inch longer than 3 inches.

Write the mixed number. The leaf is $3\frac{3}{4}$ inches long.

 Write the equation.
Three $\frac{1}{4}$-tiles can be used to model $\frac{3}{4}$.

Write 3 as 1 + 1 + 1.

$$3\frac{3}{4} = 1 + 1 + 1 + \frac{1}{4} + \frac{1}{4} + \frac{1}{4}$$

Write $\frac{3}{4}$ as a sum of unit fractions.

So, $3\frac{3}{4} = 1 + $ _____ $ + 1 + \frac{1}{4} + \frac{1}{4} + $ _____.

Guided Practice

1. Write a mixed number for the shaded model.

 _____

Talk MATH

How are fractions and mixed numbers alike? How are they different?

538 Chapter 8 Fractions

Name _____

MY Homework

Lesson 10

Mixed Numbers and Improper Fractions

Homework Helper

Need help? connectED.mcgraw-hill.com

Kelsey made 2 pitchers of lemonade. Each pitcher holds 6 cups. She poured 4 cups of lemonade from one pitcher. What fraction of the lemonade is left?

One Way Count the wholes and the parts.

If 4 cups of lemonade are poured from one pitcher, then there is 1 full pitcher left and 2 cups remaining in the other pitcher.

$1 + \frac{2}{6} = 1\frac{2}{6}$

Another Way Count the parts.

$\frac{8}{6}$

So, there are $1\frac{2}{6}$, or $\frac{8}{6}$, pitchers of lemonade left.

Practice

Write a mixed number and an improper fraction for each shaded model.

1.

2.

_____

Lesson 10 My Homework 547

Write a mixed number and an improper fraction for each model.

3.   4.

_____   _____

5. Draw a model to write $2\frac{3}{5}$ as an improper fraction.

6. Draw a model to write $\frac{30}{4}$ as a mixed number.

Brain Builders

7. **Processes & Practices 2** **Use Number Sense** Ana walked $\frac{13}{3}$ miles. How many whole miles did Ana walk? What portion of an additional mile did she walk?

_____

_____

8. There are $5\frac{4}{5}$ cups of milk left in a carton. Write $5\frac{4}{5}$ as an improper fraction. Explain what the improper fraction represents in this situation.

_____

_____

Vocabulary Check

9. Is $\frac{10}{3}$ an improper fraction? Explain.

_____

10. **Test Practice** Amelia needs $3\frac{2}{3}$ cups of sugar to make cupcakes. Which improper fraction names this amount?

Ⓐ $\frac{5}{3}$ cups   Ⓑ $\frac{8}{3}$ cups   Ⓒ $\frac{11}{3}$ cups   Ⓓ $\frac{18}{3}$ cups

Review

Chapter 8
Fractions

Vocabulary Check

1. Use the words in the word bank to label each card.

composite   denominator   equivalent fractions
improper fraction   mixed number   numerator
prime   simplest form

$\frac{4}{6}$ → [ ]
     → [ ]

$\frac{1}{6}$ → [ ]

$\frac{2}{5}$ and $\frac{4}{10}$
[ ]

$\frac{12}{4}$ → [ ]

$5\frac{2}{6}$ → [ ]

4, 8, 10 → [ ]
3, 5, 11 → [ ]

Write an example of each of the following words.

2. factor pairs _____

3. greatest common factor _____

4. least common multiple _____

5. benchmark fractions _____

My Chapter Review 549

Concept Check

Find the factor pairs of each number.

6. 52

_____ and _____

_____ and _____

_____ and _____

7. 36

_____ and _____

_____ and _____

_____ and _____

_____ and _____

_____ and _____

8. 23

_____ and _____

Tell whether each number is *prime*, *composite*, or *neither*.

9. 0 _____

10. 31 _____

11. 62 _____

Circle the two fractions that are equivalent for each set of fractions.

12. $\frac{3}{4}$ $\frac{9}{12}$ $\frac{2}{6}$

13. $\frac{4}{10}$ $\frac{4}{100}$ $\frac{40}{100}$

14. $\frac{3}{5}$ $\frac{1}{4}$ $\frac{6}{10}$

Write each fraction in simplest form. If it is in simplest form, write *simplest form*.

15. $\frac{4}{10}$

16. $\frac{3}{9}$

17. $\frac{3}{10}$

Compare. Use >, <, or =. Check your answer using fraction tiles or number lines.

18. $\frac{5}{8}$ ◯ $\frac{1}{3}$

19. $\frac{1}{5}$ ◯ $\frac{4}{6}$

20. $\frac{2}{3}$ ◯ $\frac{8}{12}$

Problem Solving

21. Mrs. Evans has 13 pictures to hang on a wall. Is there any way she can arrange the pictures in rows other than 1 × 13 or 13 × 1, so that the same number of pictures is in each row? Tell whether 13 is a composite or prime number. Explain.

22. There are $\frac{2}{8}$ cup of peanuts and $\frac{1}{4}$ cup of walnuts. Is there a greater amount of peanuts or walnuts? Explain.

Brain Builders

23. Mia has two whole bananas and one-fifth of a third banana. Write an improper fraction that represents the amount of bananas she has.

24. Write a real-world problem to compare fractions. Then solve the problem. Explain how you can check the solution.

25. Test Practice Which equation is true?

Ⓐ $2\frac{2}{3} = 2 + 2 + 3$

Ⓑ $2\frac{2}{3} = 1 + 1 + 2 + 3$

Ⓒ $2\frac{2}{3} = 1 + 1 + \frac{1}{3} + \frac{1}{3}$

Ⓓ $2\frac{2}{3} = 1 + 1 + \frac{1}{3}$

Reflect

Chapter 8
Answering the
ESSENTIAL QUESTION

Use what you learned about fractions to complete the graphic organizer.

ESSENTIAL QUESTION

How can different fractions name the same amount?

Words	Pictures	Numbers

Reflect on the **ESSENTIAL QUESTION**. Write your answer below.

_____

_____

Name _____ Date _____

Score _____

Performance Task

Charity Begins at the Market

The Country Market is a popular place for fresh vegetables and homemade goodies. Sooner or later, everyone in town comes to the Market. Three students, Ashton, Pat, and Sam, each have a table at the Market. Each student is trying to raise money for the Food Bank. They are having a friendly contest to see who can raise the most money.

Show all your work to receive full credit.

Part A

Mr. Reynolds gave a total of $48 to the students. Find all the factor pairs of 48.

Equation	Factor Pairs

Part B

Ashton has raised $\frac{8}{12}$ of her goal, Pat has raised $\frac{2}{3}$ of her goal, and Sam has raised $\frac{5}{8}$ of his goal. Show whether two of these fractions are equivalent, using the number lines below.

_____ and _____ are equivalent fractions.

Online Content at connectED.mcgraw-hill.com

Part C

Find two other fractions that are equivalent to Pat's fraction. Be sure to show all your work.

Part D

Another student has joined the friendly contest at the Market. Jesse got some money from his grandparents, so he now has $\frac{7}{16}$ of his goal. Use the benchmark fraction $\frac{1}{2}$ to compare his fraction to Sam's fraction. Use > or < to write a comparison.

_____

Part E

Jesse got another donation from his neighbors. He now has $\frac{12}{16}$ of his goal. Use the number lines above to compare his new fraction to Sam's fraction. Write a comparison using < or >.

Chapter 9 Operations with Fractions

ESSENTIAL QUESTION

How can I use operations to model real-world fractions?

Watch a video!

Let's Play Games!

Name _____

 Brain Builders

MY Chapter Project

Growing Up

1. With your partner, plant one seed in your cup of potting soil. Water it and place it where it will get enough sun.

   Type of plant: _____

   Date planted: _____

2. After your seed germinates, begin measuring it daily. Record the measurements in the table below. Use fractions and mixed numbers for your measurements.

Date	Height (in inches)

3. After your plant grows to a certain height, you and your partner will make up one addition and one subtraction problem about your plant's measurements. You will write these problems on a chart for the class to solve.

   Addition Problem: _____

   Subtraction Problem: _____

4. On a separate sheet of paper, you and your partner will solve all the problems on the chart.

Name _____

Am I Ready?

Compare. Use <, >, or =.

1. $\frac{1}{3}$ ◯ $\frac{1}{6}$
2. $\frac{4}{5}$ ◯ $\frac{8}{10}$
3. $\frac{2}{5}$ ◯ $\frac{2}{3}$
4. $\frac{2}{10}$ ◯ $\frac{10}{100}$
5. $\frac{3}{5}$ ◯ $\frac{2}{8}$
6. $\frac{1}{4}$ ◯ $\frac{2}{8}$

Write each fraction in simplest form.

7. $\frac{4}{6}$ _____
8. $\frac{6}{10}$ _____
9. $\frac{3}{12}$ _____
10. $\frac{60}{100}$ _____
11. $\frac{5}{10}$ _____
12. $\frac{4}{12}$ _____

Write an equivalent fraction.

13. $\frac{1}{3}$ _____
14. $\frac{3}{4}$ _____
15. $\frac{7}{10}$ _____
16. $\frac{3}{5}$ _____
17. $\frac{2}{10}$ _____
18. $\frac{90}{100}$ _____

19. Hannah has a garden with basil, rosemary, and parsley. One sixth of her garden has basil. One half of her garden has rosemary. One-third of her garden has parsley. Draw and label a picture of Hannah's garden.

Shade the boxes to show the problems you answered correctly.

How Did I Do? 1 2 3 4 5 6 7 8 9 10 11 12 13 14 15 16 17 18 19

Online Content at connectED.mcgraw-hill.com

Name _____

MY Math Words

Review Vocabulary

denominator   mixed number   numerator   simplest form

Making Connections

Write or draw an example and a non-example of each review vocabulary term.

Fractions

Word	Example	Non-Example
denominator		
mixed number		
numerator		
simplest form		

556 Chapter 9 Operations with Fractions

MY Vocabulary Cards

Lesson 9–1

like fractions

Ideas for Use

- Write a tally mark on each card every time you read the word in this chapter or use it in your writing. Challenge yourself to use at least 5 tally marks.

- Ask students to use the blank cards to draw or write phrases or examples that will help them with concepts like *adding like fractions* and *subtracting mixed numbers*.

Fractions that have the same denominator.

Like can mean "of the same form or kind." How does this help you remember the meaning of *like fractions*?

MY Foldable

FOLDABLES Follow the steps on the back to make your Foldable.

Subtract Fractions

$$\frac{5}{9} - \frac{3}{9} =$$

$$\frac{1}{5} + \frac{2}{5} =$$

$$5 \times \frac{1}{8} =$$

Add Fractions

Multiply Fractions

Name _____

Lesson 1
Hands On
Use Models to Add Like Fractions

ESSENTIAL QUESTION
How can I use operations to model real-world fractions?

A unit fraction has a numerator of 1. It is one of the equal parts of the whole. Fractions that have the same denominators are called **like fractions**. You can use models to add like fractions.

Build It

Abigail is playing a board game with her friends. There are 5 game pieces. There is a red piece, a blue piece, a yellow piece, a green piece, and a purple piece. What is the total fraction of the pieces that are yellow, green, or purple?

Find $\frac{1}{5} + \frac{1}{5} + \frac{1}{5}$.

1. Model three $\frac{1}{5}$-fraction tiles.

Each game piece represents a unit fraction.

2. Add the like fractions.

There are three $\frac{1}{5}$-fraction tiles all together.

So, $\frac{1}{5} + \frac{1}{5} + \frac{1}{5} = \frac{3}{5}$. $\frac{3}{5}$ is the sum of three unit fractions of $\frac{1}{5}$.

So, the total fraction of game pieces that are yellow, green, or purple is $\frac{\square}{\square}$.

Online Content at connectED.mcgraw-hill.com

Not all fractions are unit fractions. You can decompose, or break down, a fraction into a sum of unit fractions.

Try It

Find $\frac{3}{8} + \frac{2}{8}$.

 Model $\frac{3}{8}$.

$\frac{3}{8}$ is the sum of three unit fractions of $\frac{1}{8}$.

 Model $\frac{2}{8}$.

$\frac{2}{8}$ is the sum of two unit fractions of $\frac{1}{8}$.

 Add.
Count the total number of $\frac{1}{8}$-tiles. Notice that the parts being joined are part of the same whole, and each unit fraction is the same size.

There are _____ $\frac{1}{8}$-tiles.

So, $\frac{3}{8} + \frac{2}{8} = \dfrac{\Box}{\Box}$.

Talk About It

1. **Processes & Practices** **Identify Structure** The fraction $\frac{5}{8}$ can be considered as the sum of $\frac{1}{8} + \frac{1}{8} + \frac{1}{8} + \frac{1}{8} + \frac{1}{8}$. It can also be considered as the sum of $\frac{3}{8} + \frac{2}{8}$. Write two other ways to think of $\frac{5}{8}$ as a sum.

_____

_____

562 Chapter 9 Operations with Fractions

Name _____

Practice It

Model the sum using fraction tiles. Draw the model. Then add.

2. $\frac{1}{6} + \frac{4}{6} =$ _____

3. $\frac{4}{8} + \frac{2}{8} =$ _____

Use the table to answer Exercises 4 and 5.

4. The table shows the sums of several like fractions. Study the pattern in the table.

   Write a rule that you can use to add like fractions without using models.

Addends	Sum
$\frac{1}{5} + \frac{1}{5} + \frac{1}{5}$	$\frac{3}{5}$
$\frac{3}{8} + \frac{2}{8}$	$\frac{5}{8}$
$\frac{1}{10} + \frac{1}{10} + \frac{7}{10}$	$\frac{9}{10}$

_____

_____

_____

5. Use your rule from Exercise 4 to find $\frac{1}{4} + \frac{1}{4} + \frac{1}{4}$.

   _____

Algebra Write each fraction as a sum of unit fractions. Then write an equation to decompose the fraction in a different way.

6. $\frac{7}{8}$ _____

   _____

7. $\frac{5}{10}$ _____

   _____

Lesson 1 Hands On: Use Models to Add Like Fractions 563

Apply It

8. **Processes &Practices 2** **Use Number Sense** Claire walked $\frac{1}{3}$ mile on Saturday and $\frac{1}{3}$ mile on Sunday. How many miles did she walk in all?

_____

9. Miguel cut a cantaloupe into 8 slices. He ate 1 slice and his friends ate 4 slices. What fraction of the cantaloupe did they eat?

_____

10. There are 12 children at lunch. Six children are eating apples, four children are eating grapes, and two children are eating celery. What fraction of the children are eating fruit?

_____

11. **Processes &Practices 4** **Model Math** Write three different ways to decompose $\frac{9}{10}$ into a sum.

_____

_____

_____

Write About It

12. When adding like fractions, why is the numerator the only part of the fraction that changes?

_____

_____

_____

Name _____

MY Homework

Lesson 1

Hands On: Use Models to Add Like Fractions

Homework Helper   Need help? connectED.mcgraw-hill.com

Caroline has 6 cups. She fills a cup with juice for each of her 5 friends. What is the fraction of the total number of cups that Caroline gives to her friends?

Each cup is $\frac{1}{6}$ of the total number of cups.

Find $\frac{1}{6} + \frac{1}{6} + \frac{1}{6} + \frac{1}{6} + \frac{1}{6}$.

1 Use five $\frac{1}{6}$-fraction tiles to model the addition.

2 Add the like fractions.

Count the total number of $\frac{1}{6}$-fraction tiles. There are 5.

$\frac{1}{6} + \frac{1}{6} + \frac{1}{6} + \frac{1}{6} + \frac{1}{6} = \frac{5}{6}$

So, the fraction of the total number of cups that Caroline gives to her friends is $\frac{5}{6}$.

Practice

Model the sum using fraction tiles. Draw the model. Then add.

1. $\frac{1}{3} + \frac{1}{3} =$ _____

2. $\frac{2}{12} + \frac{6}{12} =$ _____

3. $\frac{3}{10} + \frac{4}{10} =$ _____

4. $\frac{5}{8} + \frac{1}{8} =$ _____

Write an addition sentence for each model. Then find the sum.

5.   6.

_____   _____

7.   8.

_____   _____

Problem Solving

9. **Processes & Practices 1** **Make Sense of Problems** Nelson got a pack of 12 pencils. He took 3 pencils to put in his desk at school and 2 pencils to put in his locker at school. What is the fraction of total pencils that Nelson took to school?

_____

10. There are 10 chicks on Ginger's farm. She has 2 chicks in one outdoor pen and 5 chicks in another. The rest of the chicks are in the barn. What is the fraction of total chicks in outdoor pens?

_____

Vocabulary Check

11. Explain what is the same about two fractions that are like fractions.

_____

_____

Lesson 2
Add Like Fractions

ESSENTIAL QUESTION
How can I use operations to model real-world fractions?

Think of adding like fractions as joining parts that refer to the same whole. To add like fractions, add the numerators and keep the same denominator.

$$\frac{1}{5} + \frac{2}{5} = \frac{3}{5}$$

$$\frac{3}{10} + \frac{4}{10} = \frac{7}{10}$$

Math in My World

Example 1

Pablo spent $\frac{2}{6}$ of an hour on a jigsaw puzzle. Conrad spent $\frac{1}{6}$ of an hour on the puzzle. How much time did they spend working on the puzzle in all?

Find $\frac{2}{6} + \frac{1}{6}$.

1. Add the numerators. Keep the same denominator.

$$\frac{2}{6} + \frac{1}{6} = \frac{2+1}{6}$$
$$= \frac{3}{6}$$

2. Write the sum in simplest form.

$$\frac{3}{6} = \frac{3 \div 3}{6 \div 3} = \frac{1}{2}$$

So, they spent $\frac{\Box}{\Box}$ hour on the puzzle.

Online Content at connectED.mcgraw-hill.com

Example 2

Find $\frac{3}{10} + \frac{5}{10}$.

1 Add.
$$\frac{3}{10} + \frac{5}{10} = \frac{3+5}{10}$$
$$= \frac{8}{10}$$

2 Simplify.
$$\frac{8}{10} = \frac{8 \div 2}{10 \div 2} = \frac{4}{5}$$

Helpful Hint
To simplify a fraction, divide the numerator and denominator by the greatest common factor.

So, $\frac{3}{10} + \frac{5}{10} = \frac{\square}{\square}$.

Check Use models to check.

$\frac{3}{10} + \frac{5}{10} = \frac{8}{10}$

$\frac{8}{10} = \frac{4}{5}$

Guided Practice

Find each sum. Write in simplest form.

1. $\frac{1}{6} + \frac{2}{6} = \frac{\square}{\square}$

2. $\frac{2}{5} + \frac{2}{5} = \frac{\square}{\square}$

Talk MATH
Describe two ways to decompose $\frac{4}{5}$ into a sum.

Name ....

Independent Practice

Find each sum. Write in simplest form.

3. $\frac{1}{3} + \frac{1}{3} =$ _____

4. $\frac{2}{6} + \frac{3}{6} =$ _____

5. $\frac{3}{8} + \frac{1}{8} =$ _____

6. $\frac{5}{8} + \frac{2}{8} =$ _____

7. $\frac{3}{10} + \frac{7}{10} =$ _____

8. $\frac{2}{5} + \frac{1}{5} =$ _____

9. $\frac{2}{12} + \frac{5}{12} =$ _____

10. $\frac{1}{4} + \frac{2}{4} =$ _____

11. $\frac{1}{6} + \frac{2}{6} =$ _____

Algebra Write each fraction as a sum of unit fractions. Then write an equation to decompose the fraction in a different way.

12. $\frac{4}{5}$ _____

_____

13. $\frac{3}{4}$ _____

_____

14. $\frac{3}{8}$ _____

_____

Lesson 2 Add Like Fractions

Problem Solving

15. Marcela ate $\frac{4}{10}$ of a box of crackers last week. Then she ate $\frac{2}{10}$ of the box of crackers this week. What fraction of the box of crackers did Marcela eat all together? Write in simplest form.

16. Greyson's mom used 4 out of 12 eggs to make pancakes. Then she used 3 out of 12 eggs to make cupcakes. What fraction of a dozen eggs did she use in all? *(Hint: 1 dozen = 12)*

Brain Builders

17.   **Model Math** Bryce has to complete 100 minutes of writing every ten days. He writes for 10 minutes each day. What fraction of his writing requirement does Bryce complete in six days? Write in simplest form.

18.   **Stop and Reflect** Write two fractions whose sum is greater than 1.

If the denominators are the same, how can you know if the sum of two fractions will be greater than 1?

19. **Building on the Essential Question** Write and solve a real-world problem that involves the sum of two like fractions.

570   Chapter 9 Operations with Fractions

Name _____

MY Homework

Lesson 2

Add Like Fractions

Homework Helper

Need help? connectED.mcgraw-hill.com

A bike rack has 10 slots. There are bikes in 2 of the slots. Then 4 more people come and park their bikes in the rack. What fraction of the bike rack is being used?

Find $\frac{2}{10} + \frac{4}{10}$.

1 Add the numerators.
$$\frac{2}{10} + \frac{4}{10} = \frac{2+4}{10}$$
$$= \frac{6}{10}$$

2 Write in simplest form.
$$\frac{6 \div 2}{10 \div 2} = \frac{3}{5}$$

So, the total fraction of the bike rack that is being used is $\frac{3}{5}$.

Practice

Find each sum. Write in simplest form.

1. $\frac{2}{6} + \frac{3}{6} =$ _____

2. $\frac{3}{8} + \frac{3}{8} =$ _____

3. $\frac{1}{4} + \frac{1}{4} =$ _____

4. $\frac{5}{12} + \frac{3}{12} =$ _____

Lesson 2 My Homework 571

Find each sum. Write in simplest form.

5. $\frac{3}{5} + \frac{1}{5} = $ _____

6. $\frac{4}{10} + \frac{1}{10} = $ _____

7. $\frac{1}{6} + \frac{3}{6} = $ _____

8. $\frac{50}{100} + \frac{30}{100} = $ _____

Brain Builders

Solve. Write the answer in simplest form.

9. **Processes &Practices 2** **Use Number Sense** It took Yuki $\frac{1}{6}$ of an hour to water her flowers and $\frac{4}{6}$ of an hour to weed her garden. What fraction of an hour did Yuki work in her yard?

_____

10. Yesterday it rained $\frac{5}{8}$ inch. Today it rained $\frac{1}{8}$ inch. What is the total amount of rain for the two days?

_____

11. Claudio walked $\frac{3}{10}$ mile in the morning. He walked the same distance in the afternoon, and again in the evening. How far did Claudio walk all together?

_____

12. **Test Practice** Pierre has 12 packages to send. He sent 1 package on Monday, 1 more package on Tuesday, and 2 more packages on Wednesday. What fraction of the packages has Pierre sent so far?

   Ⓐ $\frac{1}{4}$
   Ⓒ $\frac{8}{12}$
   Ⓑ $\frac{1}{3}$
   Ⓓ $\frac{2}{3}$

Need more practice? Download Extra Practice at connectED.mcgraw-hill.com

Name

Lesson 3
Hands On
Use Models to Subtract Like Fractions

ESSENTIAL QUESTION
How can I use operations to model real-world fractions?

You can use models to subtract like fractions. Think of subtracting fractions as separating parts that refer to the same whole.

Build It

Brayden used $\frac{4}{5}$ of his coins in the first two levels of a video game. He used $\frac{1}{5}$ of his coins in the first level. What fraction of his coins did he use in the second level?
Find $\frac{4}{5} - \frac{1}{5}$.

1 Model $\frac{4}{5}$.

$\frac{4}{5} = \frac{1}{5} + \frac{1}{5} + \frac{1}{5} + \frac{1}{5}$

2 Subtract $\frac{1}{5}$.

Remove, or separate, $\frac{1}{5}$ from $\frac{4}{5}$.

3 Count the total number of $\frac{1}{5}$-tiles that are left.

There are three $\frac{1}{5}$-fraction tiles left.

$\frac{4}{5} - \frac{1}{5} = \frac{3}{5}$ So, Brayden used $\dfrac{\square}{\square}$ of his coins in the second level.

Check using addition. $\dfrac{\square}{\square} + \dfrac{1}{5} = \dfrac{4}{5}$

Online Content at connectED.mcgraw-hill.com

Try It

Find $\frac{7}{8} - \frac{3}{8}$.

 Model $\frac{7}{8}$.

Notice that $\frac{7}{8}$ is the sum of seven unit fractions of $\frac{1}{8}$.

Separate, or remove, $\frac{3}{8}$ from $\frac{7}{8}$.

$\frac{7}{8} - \frac{3}{8}$

 Subtract $\frac{3}{8}$.

Notice that $\frac{3}{8}$ is the sum of three unit fractions of $\frac{1}{8}$.

 Count the remaining $\frac{1}{8}$-tiles.

There are ____ remaining $\frac{1}{8}$-tiles.

$\frac{7}{8} - \frac{3}{8} = \frac{\square}{\square}$.

Simplify. $\frac{4 \div 4}{8 \div 4} = \frac{\square}{2}$

So, $\frac{7}{8} - \frac{3}{8} = \frac{\square}{\square}$.

Check using addition. $\frac{\square}{\square} + \frac{3}{8} = \frac{7}{8}$

Talk About It

1. **Processes & Practices** ➡ **Draw a Conclusion** What do you notice about the denominators when subtracting like fractions?

_____

_____

Name _____

Practice It

Model the difference using fraction tiles. Then subtract.

2. $\dfrac{4}{6} - \dfrac{3}{6} =$ _____

3. $\dfrac{9}{12} - \dfrac{4}{12} =$ _____

4. $\dfrac{6}{10} - \dfrac{3}{10} =$ _____

5. $\dfrac{4}{5} - \dfrac{2}{5} =$ _____

Use the table for Exercises 6 and 7.

6. The table shows the difference of several like fractions. Study the pattern in the table. Write a rule that you can use to subtract like fractions without using models.

Like Fractions	Difference
$\dfrac{4}{5} - \dfrac{1}{5}$	$\dfrac{3}{5}$
$\dfrac{7}{8} - \dfrac{3}{8}$	$\dfrac{4}{8}$ or $\dfrac{1}{2}$
$\dfrac{9}{10} - \dfrac{6}{10}$	$\dfrac{3}{10}$

_____

_____

_____

7. Use your rule from Exercise 6 to find $\dfrac{11}{12} - \dfrac{8}{12}$. _____

Algebra Find each unknown. Use fraction tiles.

8. $\dfrac{2}{4} - \dfrac{1}{4} = y$

   $y =$ _____

9. $\dfrac{5}{8} - \dfrac{2}{8} = x$

   $x =$ _____

10. $\dfrac{7}{10} - \dfrac{3}{10} = t$

    $t =$ _____

11. $\dfrac{5}{6} - \dfrac{4}{6} = s$

    $s =$ _____

Lesson 3 Hands On: Use Models to Subtract Like Fractions

Apply It

12. Ann ate 2 slices of a pizza and Teresa ate 3 slices of the pizza. The pizza had 8 slices. What is the difference in the amount of pizza that they ate written as a fraction?

_____

13. **Processes &Practices** **Use Symbols** Virginia had $\frac{8}{12}$ pictures left on her memory card. Then she took $\frac{7}{12}$ of the pictures. What fraction of the pictures does she have left? Write an equation to solve.

_____

14. **Processes &Practices** **Reason** Draw a model to find the unknown.
$\frac{4}{5} - \frac{1}{5} = ?$

Write About It

15. How is subtracting like fractions similar to adding like fractions?

_____

_____

Name

MY Homework

Lesson 3

Hands On: Use Models to Subtract Like Fractions

Homework Helper

Need help? connectED.mcgraw-hill.com

Nicole has 8 charms. She put 7 on her bracelet. At the park, 2 charms fell off the bracelet. What fraction of Nicole's charms are on her bracelet now?

Find $\frac{7}{8} - \frac{2}{8}$.

1 Model $\frac{7}{8}$.

$\frac{7}{8} = \frac{1}{8} + \frac{1}{8} + \frac{1}{8} + \frac{1}{8} + \frac{1}{8} + \frac{1}{8} + \frac{1}{8}$

2 Subtract $\frac{2}{8}$.

Remove, or separate, $\frac{2}{8}$ from $\frac{7}{8}$.

3 Count the total number of $\frac{1}{8}$-tiles that are left.

There are five $\frac{1}{8}$-fraction tiles left.

$\frac{7}{8} - \frac{2}{8} = \frac{5}{8}$

So, Nicole has $\frac{5}{8}$ of her charms left on her bracelet.

Check

$\frac{5}{8} + \frac{2}{8} = \frac{7}{8}$

So, the answer is correct.

Lesson 3 My Homework 577

Practice

Model the difference using fraction tiles. Then subtract.

1. $\frac{5}{6} - \frac{1}{6} =$ _____

2. $\frac{2}{3} - \frac{1}{3} =$ _____

3. $\frac{9}{10} - \frac{6}{10} =$ _____

4. $\frac{6}{8} - \frac{4}{8} =$ _____

Algebra Write a subtraction equation for each model. Then subtract.

5.

6.

_____ _____

7.

8.

_____ _____

Problem Solving

9. For Friday night's play performance, $\frac{7}{8}$ of the theater was full. On Saturday night, the theater was only $\frac{5}{8}$ full. How much more of the theater was filled on Friday night than on Saturday night? Write in simplest form.

_____

10. **Processes & Practices** 2 **Use Number Sense** Liam uses $\frac{3}{4}$ of a cup of butter in his cookie recipe. Gloria uses $\frac{2}{4}$ of a cup of butter in her cookie recipe. How much more butter does Liam use than Gloria?

_____

578 Chapter 9 Operations with Fractions

Name _____

Lesson 4
Subtract Like Fractions

ESSENTIAL QUESTION
How can I use operations to model real-world fractions?

Think of subtracting like fractions as separating parts that refer to the same whole. To subtract like fractions, subtract the numerators and keep the same denominator.

$$\frac{4}{5} - \frac{2}{5} = \frac{2}{5}$$

$$\frac{8}{10} - \frac{5}{10} = \frac{3}{10}$$

Math in My World

Example 1

Liliana jogged $\frac{5}{8}$ of a mile on Monday and $\frac{3}{8}$ of a mile on Tuesday. How much farther did she jog on Monday?

Find $\frac{5}{8} - \frac{3}{8}$.

 Subtract the numerators.
Keep the same denominator.

$$\frac{5}{8} - \frac{3}{8} = \frac{5-3}{8}$$
$$= \frac{2}{8}$$

 Write the difference in simplest form.

$$\frac{2}{8} = \frac{2 \div 2}{8 \div 2} = \frac{1}{4}$$

So, Liliana jogged ⬜/⬜ mile farther on Monday.

Online Content at connectED.mcgraw-hill.com

Lesson 4 579

Example 2

Find $\frac{7}{10} - \frac{5}{10}$.

You can also use number lines to subtract like fractions. Count back five groups of $\frac{1}{10}$ to remove $\frac{5}{10}$ from $\frac{7}{10}$.

 Subtract.

$$\frac{7}{10} - \frac{5}{10} = \frac{7-5}{10}$$
$$= \frac{2}{10}$$

 Simplify.

$$\frac{2}{10} = \frac{2 \div 2}{10 \div 2} = \frac{1}{5}$$

So, $\frac{7}{10} - \frac{5}{10} = \frac{\square}{\square}$.

Guided Practice

Find each difference. Write in simplest form.

1. $\frac{6}{8} - \frac{3}{8} = \frac{\square}{\square}$

2. $\frac{11}{12} - \frac{5}{12} = \frac{\square}{\square}$

Talk MATH

Explain how to find $\frac{7}{8} - \frac{1}{8}$.

Name ....

Independent Practice

Find each difference. Write in simplest form.

3. $\dfrac{7}{8} - \dfrac{2}{8} =$ _____

4. $\dfrac{5}{6} - \dfrac{4}{6} =$ _____

5. $\dfrac{9}{10} - \dfrac{6}{10} =$ _____

6. $\dfrac{2}{3} - \dfrac{1}{3} =$ _____

7. $\dfrac{4}{5} - \dfrac{2}{5} =$ _____

8. $\dfrac{4}{6} - \dfrac{2}{6} =$ _____

9. $\dfrac{9}{12} - \dfrac{5}{12} =$ _____

10. $\dfrac{7}{10} - \dfrac{3}{10} =$ _____

11. $\dfrac{6}{10} - \dfrac{1}{10} =$ _____

Algebra Find each unknown. Match each equation to its difference in simplest form.

12. $\dfrac{4}{5} - \dfrac{1}{5} = \blacksquare$

- $\dfrac{4}{12}$

- $\dfrac{3}{5}$

13. $\dfrac{6}{12} - \dfrac{2}{12} = \blacksquare$

- $\dfrac{5}{10}$

- $\dfrac{1}{2}$

14. $\dfrac{7}{10} - \dfrac{2}{10} = \blacksquare$

- $\dfrac{1}{3}$

- $\dfrac{6}{10}$

Lesson 4 Subtract Like Fractions

Problem Solving

15. Kenji's dog ate $\frac{3}{12}$ of the treats in a box of dog treats on Monday. His dog ate $\frac{2}{12}$ of the treats on Tuesday. What fraction more of the box of the treats did his dog eat on Monday than on Tuesday?

_____

16. **Processes & Practices 4** **Model Math** Dan used $\frac{4}{10}$ of a pack of golf balls on Saturday. Then he used $\frac{2}{10}$ of the pack of golf balls on Sunday. What fraction more of the pack of golf balls did Dan use on Saturday than on Sunday? Draw a model to solve.

_____

Brain Builders

17. **Processes & Practices 3** **Find the Error** Bella has $\frac{5}{8}$ of her room left to paint. She says that if she paints $\frac{2}{8}$ of it today she will have $\frac{7}{8}$ of it left to paint tomorrow. Find and correct her mistake.

_____
_____

18. **Building on the Essential Question** When subtracting like fractions, what happens to the numerator and the denominator? Use a model to explain your reasoning.

_____
_____
_____

Name

MY Homework

Lesson 4

Subtract Like Fractions

Homework Helper

Need help? connectED.mcgraw-hill.com

It takes Margaret $\frac{5}{6}$ of an hour to do her chores. It takes Holly $\frac{3}{6}$ of an hour to do her chores. How much longer does it take Margaret to do her chores than it takes Holly?

Find $\frac{5}{6} - \frac{3}{6}$.

 Find the difference between the numerators.

$$\frac{5}{6} - \frac{3}{6} = \frac{5-3}{6}$$
$$= \frac{2}{6}$$

Write in simplest form.

$$\frac{2 \div 2}{6 \div 2} = \frac{1}{3}$$

So, it takes Margaret $\frac{1}{3}$ of an hour longer to do her chores.

Practice

Find each difference. Write in simplest form.

1. $\frac{7}{10} - \frac{4}{10} =$ _____

2. $\frac{10}{12} - \frac{3}{12} =$ _____

3. $\frac{4}{5} - \frac{3}{5} =$ _____

4. $\frac{6}{8} - \frac{4}{8} =$ _____

Find each difference. Write in simplest form.

5. $\frac{6}{8} - \frac{2}{8} =$ _____

6. $\frac{4}{10} - \frac{2}{10} =$ _____

7. $\frac{9}{12} - \frac{6}{12} =$ _____

8. $\frac{80}{100} - \frac{20}{100} =$ _____

Problem Solving

Solve. Write the answer in simplest form.

9. A beetle is $\frac{1}{5}$ inch wide and $\frac{2}{5}$ inch long. How much greater is the beetle's length than its width?

_____

10. Last Friday, $\frac{7}{10}$ of the rooms at a motel were rented. This Friday, $\frac{9}{10}$ of the rooms are rented. What fraction more of the rooms are rented this Friday than were rented last Friday?

_____

Brain Builders

11. **Processes &Practices** 2 **Use Number Sense** Denise teaches obedience classes. Last session, $\frac{11}{12}$ of the dogs passed her class. This session, $\frac{9}{12}$ of the dogs passed her class. What fraction more of the dogs passed her class last session? Write an equation to represent the situation. Write the answer in simplest form.

_____

12. **Test Practice** During basketball practice Andy made $\frac{4}{8}$ of his shots. During a game he made $\frac{2}{8}$ of his shots. What fraction more of his shots did Andy make during practice?

Ⓐ $\frac{3}{4}$

Ⓒ $\frac{2}{4}$

Ⓑ $\frac{4}{8}$

Ⓓ $\frac{1}{4}$

Check My Progress

Vocabulary Check

1. Circle the pair of **like fractions**.

   | $\frac{4}{6}$ and $\frac{5}{6}$ | $\frac{1}{2}$ and $\frac{3}{4}$ | $\frac{1}{3}$ and $\frac{2}{6}$ | $\frac{1}{10}$ and $\frac{2}{100}$ |

   Explain how you knew which pair of fractions to circle.

   _____

   _____

Concept Check

Algebra Write each fraction as a sum of unit fractions. Then write an equation to decompose the fraction in a different way.

2. $\frac{4}{5}$ _____

   _____

3. $\frac{3}{8}$ _____

   _____

Find each sum. Write in simplest form.

4. $\frac{2}{5} + \frac{2}{5} = $ _____

5. $\frac{1}{8} + \frac{5}{8} = $ _____

6. $\frac{2}{6} + \frac{1}{6} = $ _____

Find each difference. Write in simplest form.

7. $\frac{7}{10} - \frac{4}{10} = $ _____

8. $\frac{3}{4} - \frac{1}{4} = $ _____

9. $\frac{10}{12} - \frac{6}{12} = $ _____

Problem Solving

10. Elizabeth has $\frac{7}{10}$ of a dollar. If she spends $\frac{4}{10}$ of a dollar, what fraction of a dollar will she have left?

    _____

11. A game uses tokens. Two-sixths of the tokens are blue. Three-sixths of the tokens are gold. What fraction of the tokens are either blue or gold?

    _____

Brain Builders

12. A glass has $\frac{6}{8}$ cup of water. If Jack pours out $\frac{3}{8}$ cup of the water, how much water will be left? Draw a model to help solve.

    _____

13. A recipe uses $\frac{2}{8}$ cup of apple juice, $\frac{1}{8}$ cup of grape juice, and $\frac{4}{8}$ cup of orange juice. What is the total amount of apple juice, grape juice, and orange juice in the recipe?

    _____

14. **Test Practice** What is the sum of $\frac{2}{10}$ and $\frac{3}{10}$ in simplest form?

    Ⓐ $\frac{5}{20}$ Ⓒ $\frac{23}{10}$

    Ⓑ $\frac{1}{10}$ Ⓓ $\frac{1}{2}$

586 Chapter 9 Operations with Fractions

Name _____

Lesson 5
Problem-Solving Investigation
STRATEGY: Work Backward

ESSENTIAL QUESTION
How can I use operations to model real-world fractions?

Learn the Strategy

Nathan used some flour for a cake recipe. He used $\frac{1}{4}$ of the bag of flour for a bread recipe. There is $\frac{2}{4}$ of the bag left. What fraction of the bag did Nathan use for the cake?

1 Understand

What facts do you know?

Nathan used some flour for a cake recipe and _____ of the bag for a bread recipe. He has _____ of the bag left.

What do you need to find?

Find the fraction of the bag of flour that was used for the _____.

2 Plan

I will work backward to solve the problem.

3 Solve

| $\frac{1}{4}$ | $\frac{1}{4}$ | $\frac{1}{4}$ | $\frac{1}{4}$ |
| bread | flour left | | cake |

$\frac{1}{4} + \frac{1}{4} + \frac{1}{4} + \frac{1}{4} = 1$ whole bag of flour

So, _____ of the bag of flour was used for the cake.

4 Check

Does your answer make sense? Explain.

Online Content at connectED.mcgraw-hill.com

Lesson 5 587

Practice the Strategy

Lily, Noah, and Kaylee are sharing pizza. Noah ate $\frac{2}{6}$ of the pizza. Kaylee ate $\frac{1}{6}$ of the pizza. There is $\frac{2}{6}$ of the pizza left. What fraction of the pizza did Lily eat?

1 Understand

What facts do you know?

_____
_____

What do you need to find?

_____

2 Plan

_____
_____

3 Solve

4 Check

Does your answer make sense? Explain.

_____
_____

Name ....

Apply the Strategy

Solve each problem by working backward.

1. **Processes & Practices 1** **Make Sense of Problems** Chloe did some of her homework before dinner. She did $\frac{2}{6}$ of her homework after dinner. She has $\frac{1}{6}$ of her homework left. What fraction of her homework did Chloe do before her dinner? Write in simplest form.

   _____

Brain Builders

2. A bag of marbles contains red, green, and blue marbles. One-sixth of the marbles are red. Two-sixths of the marbles are green. James says that there are fewer blue marbles in the bag than green. Do you agree or disagree with James? Explain.

   _____

   _____

3. Brandi and her mom are at a pet store. The pet store has 12 reptiles. Of the reptiles, $\frac{2}{12}$ are snakes and the rest are lizards or turtles. If there are the same number of lizards and turtles, what fraction of the reptiles are lizards?

   _____

Lesson 5 Problem-Solving Investigation 589

Review the Strategies

Use any strategy to solve each problem.
- Work backward.
- Use logical reasoning.
- Look for a pattern.
- Make a model.

4. There are 16 books on a shelf. Four of the books are about animals. Two are adventure. The rest are mystery. What fraction of the books are mystery books?

5. There are 10 pieces of chalk. Two tenths of the chalk is pink. One piece is blue. The rest are white. What fraction of the pieces of chalk are white?

6. Giselle played with some friends on Monday. She played with 2 times as many friends on Wednesday. This was 4 more than on Friday. On Friday, she played with 4 friends. How many did she play with on Monday?

7. **Processes & Practices 5** **Use Math Tools** Mrs. Vargas is making costumes for a play. She needs 3 buttons for each costume. Complete the table to find how many buttons she will need for 22 costumes.

Costumes	Buttons
8	24
9	27
10	30
12	36
13	39
20	60
21	
22	

Name _____

MY Homework

Lesson 5

Problem Solving:
Work Backward

Homework Helper

Need help? connectED.mcgraw-hill.com

Vincent decided to give away his marble collection. He gave some of the marbles to Sam. He gave $\frac{2}{8}$ of the marbles to Jenny, and he gave $\frac{3}{8}$ of the marbles to Molly. What fraction of the marbles did Vincent give to Sam?

1 Understand

What facts do you know?

Vincent gave $\frac{2}{8}$ of his marbles to Jenny and $\frac{3}{8}$ of his marbles to Molly.

What do you need to find?

I need to find the fraction of marbles Vincent gave to Sam.

2 Plan

I will work backward to solve the problem.

3 Solve

Jenny got $\frac{2}{8}$ of the marbles.

Molly got $\frac{3}{8}$ of the marbles.

That leaves another $\frac{3}{8}$.

So, Sam got $\frac{3}{8}$ of the marbles.

4 Check

Does the answer make sense?

$\frac{2}{8} + \frac{3}{8} + \frac{3}{8} = \frac{8}{8}$ or 1   So, the answer is reasonable.

Lesson 5 My Homework 591

Problem Solving

Solve each problem by working backward.

1. Marla, Jamie, and Sarah have each led their book club's monthly meeting. Jamie has led $\frac{2}{6}$ of the meetings, and Sarah has led $\frac{1}{6}$ of the meetings. What fraction of the meetings has Marla led?

2. Suzanne dropped her penny jar. She found some of the pennies, but some are still missing. She found $\frac{6}{10}$ of the pennies on the rug. She found $\frac{3}{10}$ of the pennies on the couch. What fraction of pennies is still missing?

Brain Builders

3. **Processes & Practices 2** **Use Number Sense** Noah spent some of his allowance on Monday, $\frac{1}{8}$ of his allowance on Tuesday, $\frac{1}{8}$ of it on Wednesday, and $\frac{2}{8}$ of it on Thursday. Noah has none of his allowance money left. What fraction of his allowance did he spend on Monday? Explain.

Lesson 6
Add Mixed Numbers

ESSENTIAL QUESTION
How can I use operations to model real-world fractions?

Mixed numbers are numbers with a whole number and a fraction. You can decompose mixed numbers to add them. Use the Associative Property to group the whole numbers and like fractions together.

Math in My World

Example 1

Madison made a fruit salad. She used $3\frac{1}{4}$ cups of strawberries and $2\frac{1}{4}$ cups of blueberries. How many cups of berries did Madison use all together?

Find $3\frac{1}{4} + 2\frac{1}{4}$.

Decompose each mixed number as a sum of whole numbers and unit fractions.

$3\frac{1}{4} + 2\frac{1}{4} = 1 + 1 + 1 + \frac{1}{4} + 1 + 1 + \frac{1}{4}$ Write as a sum of whole numbers and like fractions.

$= (1 + 1 + 1 + 1 + 1) + \left(\frac{1}{4} + \frac{1}{4}\right)$ Associative Property

$= \underline{} + \frac{1}{4} + \frac{1}{4}$ Add the wholes. There are 5 wholes.

$= 5 + \frac{2}{4}$ Add the like fractions. $\frac{1}{4} + \frac{1}{4} = \frac{2}{4}$

$= 5\frac{\square}{\square}$ Simplify.

So, Madison used $5\frac{\square}{\square}$ cups of berries.

Check

Use models. $5 + \frac{2}{4} = 5\frac{1}{2}$

You can also write each mixed number as an equivalent improper fraction.

Example 2

Find $1\frac{1}{3} + 2\frac{1}{3}$.

  Write each mixed number as an equivalent improper fraction.

$1\frac{1}{3} = \frac{3}{3} + \frac{1}{3} = \frac{3+1}{3} = \frac{\square}{\square}$

$2\frac{1}{3} = \frac{3}{3} + \frac{3}{3} + \frac{1}{3}$

$= \frac{3+3+1}{3} = \frac{\square}{\square}$

Helpful Hint
To add like fractions, add the numerators and keep the denominator the same.

 Add like fractions.

$\frac{4}{3} + \frac{7}{3} = \frac{4+7}{3} = \frac{\square}{\square}$

 Simplify. Write the improper fraction as a mixed number. The model shows 11 divided into groups of 3.

$\frac{11}{3} = 3\frac{2}{3}$

So, $1\frac{1}{3} + 2\frac{1}{3} = \square\frac{\square}{\square}$

Guided Practice

1. Find the sum. Write in simplest form.

$2\frac{3}{6} + 2\frac{1}{6} = \underline{\quad} + \underline{\quad} + \frac{3}{6} + \underline{\quad} + \underline{\quad} + \frac{1}{6}$

$= (\underline{\quad} + \underline{\quad} + \underline{\quad}) + \left(\frac{3}{6} + \frac{1}{6}\right)$

$= 4 + \frac{\square}{\square}$, or $\square\frac{\square}{\square}$

Talk MATH

Explain how adding mixed numbers is different than adding whole numbers.

594 Chapter 9 Operations with Fractions

Independent Practice

Find each sum. Write in simplest form. Use fraction models to check.

2. $2\frac{1}{3} + 1\frac{1}{3} =$ _____

3. $5\frac{1}{8} + 2\frac{3}{8} =$ _____

4. $5\frac{1}{4} + 5\frac{1}{4} =$ _____

5. $4\frac{1}{5} + 4\frac{3}{5} =$ _____

6. $4\frac{3}{8} + 2\frac{4}{8} =$ _____

7. $6\frac{2}{6} + 1\frac{1}{6} =$ _____

8. $3\frac{1}{10} + 1\frac{7}{10} =$ _____

9. $1\frac{2}{12} + 7\frac{2}{12} =$ _____

10. $3\frac{2}{8} + 2\frac{2}{8} =$ _____

Circle the sum that could *not* represent each mixed number.

11. $2\frac{2}{5}$ $1 + 1 + \frac{2}{5}$ $1 + 1 + \frac{1}{5} + \frac{1}{5}$ $2 + \frac{1}{5}$

12. $1\frac{3}{4}$ $1 + \frac{1}{4}$ $1 + \frac{1}{4} + \frac{1}{4} + \frac{1}{4}$ $\frac{4}{4} + \frac{2}{4} + \frac{1}{4}$

13. $3\frac{1}{8}$ $1 + 1 + 1 + \frac{1}{8}$ $\frac{8}{8} + \frac{8}{8} + \frac{1}{8}$ $\frac{8}{8} + \frac{8}{8} + \frac{8}{8} + \frac{1}{8}$

14. $4\frac{1}{2}$ $\frac{2}{2} + \frac{2}{2} + \frac{1}{2}$ $1 + 1 + 1 + 1 + \frac{1}{2}$ $\frac{2}{2} + \frac{2}{2} + \frac{2}{2} + \frac{2}{2} + \frac{1}{2}$

Lesson 6 Add Mixed Numbers

Problem Solving

15. Brennan ate $2\frac{1}{4}$ apples. Then he ate another $1\frac{2}{4}$ apples the next day. How many apples did Brennan eat all together? Write in simplest form.

_____

16. The bakery shop purchased $7\frac{2}{10}$ pounds of sugar and $7\frac{5}{10}$ pounds of flour. How much sugar and flour did they purchase in all?

_____

Brain Builders

17. **Processes &Practices 2** **Use Number Sense** Mrs. Argo has $3\frac{7}{12}$ boxes of pens. She has $4\frac{1}{12}$ boxes of pencils and 2 boxes of markers. How many boxes of pens, pencils, and markers does she have? Write in simplest form.

_____

18. **Processes &Practices 4** **Model Math** Write a real-world problem using mixed numbers whose sum is $1\frac{3}{5}$. Include a model.

_____

_____

19. **Building on the Essential Question** How can a mixed number be written as a sum?

_____

_____

_____

Name _____

Lesson 6

Add Mixed Numbers

Homework Helper

Need help? connectED.mcgraw-hill.com

Gavin put $2\frac{1}{3}$ scoops of chili in his bowl. He put $4\frac{1}{3}$ scoops of chili in his dad's bowl. How many scoops of chili do Gavin and his dad have in all?

Find $2\frac{1}{3} + 4\frac{1}{3}$.

1 Change the mixed numbers into improper fractions.

$2\frac{1}{3} = \frac{3}{3} + \frac{3}{3} + \frac{1}{3} = \frac{3+3+1}{3} = \frac{7}{3}$

$4\frac{1}{3} = \frac{3}{3} + \frac{3}{3} + \frac{3}{3} + \frac{3}{3} + \frac{1}{3} = \frac{3+3+3+3+1}{3} = \frac{13}{3}$

2 Add the like fractions.

$\frac{7}{3} + \frac{13}{3} = \frac{20}{3}$

3 Simplify.

$\frac{20}{3} = 6\frac{2}{3}$

$\frac{3}{3} + \frac{3}{3} + \frac{3}{3} + \frac{3}{3} + \frac{3}{3} + \frac{3}{3} + \frac{2}{3} = \frac{20}{3}$ or $6\frac{2}{3}$

So, Gavin and his dad have $6\frac{2}{3}$ scoops of chili in all.

Practice

Find each sum. Write in simplest form.

1. $4\frac{1}{4} + 2\frac{2}{4} =$ _____

2. $3\frac{3}{6} + 6\frac{1}{6} =$ _____

Lesson 6 My Homework 597

Find each sum. Write in simplest form.

3. $6\frac{2}{5} + 3\frac{2}{5} =$ _____

4. $4\frac{1}{6} + 1\frac{2}{6} =$ _____

5. $2\frac{1}{4} + 9\frac{1}{4} =$ _____

6. $7\frac{4}{8} + 1\frac{3}{8} =$ _____

7. $5\frac{6}{10} + 8\frac{3}{10} =$ _____

8. $12\frac{5}{10} + 6\frac{1}{10} =$ _____

Problem Solving

Solve. Write the answer in simplest form.

9. James cut $1\frac{1}{4}$ dozen flowers for a bouquet. Gwen added $1\frac{2}{4}$ dozen flowers to the bouquet. How many dozen flowers are there all together?

_____

10. On Monday, Simon's class filled $3\frac{2}{5}$ boxes with books to donate to charity. On Wednesday, the class filled $4\frac{2}{5}$ more boxes with books to donate. How many boxes of books will Simon's class donate in all?

_____

Brain Builders

11. **Processes &Practices 2** **Use Number Sense** Marissa rode her bike to the park and back home. Halfway to the park, she stopped at the library. She lives $1\frac{3}{10}$ miles from the library. How many miles did Marissa ride her bike in all?

_____

12. **Test Practice** Nate is $10\frac{9}{12}$ years old. How old will he be in $2\frac{5}{12}$ more years?

   Ⓐ $13\frac{1}{6}$ years old   Ⓑ $13\frac{1}{3}$ years old   Ⓒ $12\frac{1}{6}$ years old   Ⓓ $12\frac{4}{12}$ years old

Name _____

Lesson 7
Subtract Mixed Numbers

ESSENTIAL QUESTION
How can I use operations to model real-world fractions?

You can use equivalent fractions to subtract mixed numbers.

 Math in My World

Example 1

Olivia had $4\frac{3}{4}$ boxes of games. She gave $1\frac{1}{4}$ of the boxes to her brother. How many boxes of games does she have left?

Find $4\frac{3}{4} - 1\frac{1}{4}$.

1 Write each mixed number as an equivalent improper fraction.

$4\frac{3}{4} =$ $+$ $+$ $+ \frac{4}{4} + \frac{3}{4} = \frac{4+4+4+4+3}{4} = \dfrac{\Box}{\Box}$

$1\frac{1}{4} =$ $+ \frac{1}{4} = \frac{4+1}{4} = \dfrac{\Box}{\Box}$

2 Subtract like fractions.

$\frac{19}{4} - \frac{5}{4} = \frac{19-5}{4} = \dfrac{\Box}{\Box}$

3 Simplify. The model below shows 14 divided into groups of 4.

$\frac{14}{4} = 3\frac{2}{4} = 3\dfrac{\Box}{\Box}$

So, Olivia has _____ boxes of games left.

Check Use addition to check.
$3\frac{2}{4} + 1\frac{1}{4} = 1 + 1 + 1 + \frac{2}{4} + 1 + \frac{1}{4}$
$= 4\frac{3}{4}$

Online Content at connectED.mcgraw-hill.com

Example 2

Find $3\frac{1}{5} - 2\frac{2}{5}$.

 Write each mixed number as an equivalent improper fraction.

$$3\frac{1}{5} = \frac{5}{5} + \frac{5}{5} + \frac{5}{5} + \frac{1}{5} = \frac{5+5+5+1}{5} = \frac{\square}{\square}$$

$$2\frac{2}{5} = \frac{5}{5} + \frac{5}{5} + \frac{2}{5} = \frac{5+5+2}{5} = \frac{\square}{\square}$$

 Subtract like fractions.

$$\frac{16}{5} - \frac{12}{5} = \frac{16-12}{5} = \frac{\square}{\square}$$ This fraction is already in simplest form.

So, $3\frac{1}{5} - 2\frac{2}{5} = \frac{\square}{\square}$.

Check Use addition to check.

$$\frac{4}{5} + 2\frac{2}{5} = \frac{4}{5} + \underline{\qquad} + \underline{\qquad} + \frac{2}{5}$$

$$= 2 + \frac{6}{5}$$

$$= 2 + \frac{5}{5} + \frac{\square}{\square}$$

$$= \underline{\qquad}$$

Talk MATH

Refer to the addition check in Example 2. What property allows you to group the whole numbers together and the like fractions together when adding?

Guided Practice

Find each difference. Write in simplest form.

1. $5\frac{4}{8} - 1\frac{1}{8} = $ _____

2. $6\frac{1}{3} - 2\frac{2}{3} = $ _____

Name _____

Independent Practice

Find each difference. Write in simplest form. Use addition to check.

3. $5\frac{4}{5} - 2\frac{2}{5} =$ _____

4. $3\frac{4}{6} - 1\frac{5}{6} =$ _____

5. $8\frac{2}{3} - 3\frac{2}{3} =$ _____

6. $6\frac{1}{4} - 4\frac{2}{4} =$ _____

7. $4\frac{2}{3} - 2\frac{1}{3} =$ _____

8. $6\frac{3}{4} - 3\frac{2}{4} =$ _____

9. $9\frac{5}{6} - 7\frac{2}{6} =$ _____

10. $6\frac{7}{8} - 1\frac{5}{8} =$ _____

11. $4\frac{6}{12} - 3\frac{5}{12} =$ _____

Lesson 7 Subtract Mixed Numbers

Problem Solving

12. **Processes & Practices 7** **Identify Structure** Mrs. Garcia has $5\frac{6}{8}$ boxes of paper. She uses $3\frac{7}{8}$ boxes. How many boxes does she have left? Write in simplest form.

13. Erwin had $2\frac{5}{12}$ gallons of lemonade. He spilled $\frac{11}{12}$ gallon of lemonade. How much lemonade does he have left? Write in simplest form.

Brain Builders

14. **Processes & Practices 7** **Identify Structure** Abby had $4\frac{7}{10}$ boxes of crackers. She used $1\frac{9}{10}$ boxes of crackers to make a snack for her class. How many boxes of crackers does she have left? Write in simplest form.

15. **Processes & Practices 1** **Keep Trying** A number minus $2\frac{1}{3}$ equals $3\frac{1}{3}$. What is the number minus $1\frac{1}{3}$?

16. **? Building on the Essential Question** Explain how you can check the answer to a subtraction problem involving mixed numbers. Include an example to support your reasoning.

602 Chapter 9 Operations with Fractions

MY Homework

Lesson 7
Subtract Mixed Numbers

Homework Helper

Need help? connectED.mcgraw-hill.com

Joanna has $4\frac{1}{6}$ inches of ribbon. She cuts off $1\frac{5}{6}$ inches of the ribbon that are frayed. How long is Joanna's ribbon now?

Find $4\frac{1}{6} - 1\frac{5}{6}$.

1 Write each mixed number as an equivalent improper fraction.

$4\frac{1}{6} = \frac{6}{6} + \frac{6}{6} + \frac{6}{6} + \frac{6}{6} + \frac{1}{6} = \frac{6+6+6+6+1}{6} = \frac{25}{6}$

$1\frac{5}{6} = \frac{6}{6} + \frac{5}{6} = \frac{6+5}{6} = \frac{11}{6}$

2 Subtract like fractions.

$\frac{25}{6} - \frac{11}{6} = \frac{14}{6}$

3 Simplify.

$\frac{14}{6} = 2\frac{1}{3}$

$\frac{6}{6} + \frac{6}{6} + \frac{2}{6} = \frac{14}{6}$ or $2\frac{1}{3}$

So, Joanna's ribbon is now $2\frac{1}{3}$ inches long.

Practice

Find each difference. Write in simplest form.

1. $7\frac{5}{8} - 4\frac{2}{8} = $ _____

2. $3\frac{1}{4} - 1\frac{3}{4} = $ _____

3. $6\frac{4}{5} - 2\frac{1}{5} = $ _____

4. $8\frac{4}{6} - 2\frac{5}{6} = $ _____

Find each difference. Write in simplest form.

5. $10\frac{2}{6} - 7\frac{3}{6} = $ _____

6. $7\frac{3}{10} - 5\frac{5}{10} = $ _____

7. $15\frac{2}{3} - 8\frac{1}{3} = $ _____

8. $5\frac{2}{5} - 3\frac{4}{5} = $ _____

Problem Solving

Solve. Write the answer in simplest form.

9. **Processes & Practices 2** Use Number Sense On Saturday Rhonda filled $3\frac{1}{3}$ buckets with shells. On Sunday she filled $4\frac{2}{3}$ buckets with shells. How many more buckets of shells did Rhonda fill on Sunday?

10. Cooper worked $6\frac{1}{6}$ hours at the book fair. Amanda worked $4\frac{5}{6}$ hours at the book fair. How much longer did Cooper work?

Brain Builders

11. Sandy brought $6\frac{1}{4}$ dozen cookies to the bake sale. Alejandro brought $1\frac{3}{4}$ dozen fewer cookies than Sandy brought. Pierre brought $2\frac{3}{4}$ dozen fewer cookies than Alejandro brought. How many dozen cookies did Pierre bring?

12. **Test Practice** Tara read $4\frac{5}{8}$ pages of the arts section in the newspaper. She read $3\frac{7}{8}$ pages of the sports section. How much more of the arts section did Tara read?

Ⓐ $\frac{2}{8}$ page

Ⓒ $\frac{3}{4}$ page

Ⓑ $1\frac{2}{8}$ pages

Ⓓ $1\frac{7}{8}$ pages

Check My Progress

Vocabulary Check

Draw lines to match each of the following with its correct description or example.

1. like fractions • the form a fraction is written in when the numerator and denominator have no common factor other than 1

2. mixed number • fractions with the same denominator

3. simplest form • $5\frac{3}{4}$

Concept Check

Find each sum. Write in simplest form.

4. $1\frac{5}{10} + \frac{3}{10} =$ _____

5. $8\frac{8}{12} + 1\frac{1}{12} =$ _____

6. $5\frac{1}{4} + 3\frac{1}{4} =$ _____

7. $7\frac{20}{100} + 2\frac{40}{100} =$ _____

Find each difference. Write in simplest form.

8. $5\frac{7}{8} - 3\frac{2}{8} =$ _____

9. $7\frac{2}{3} - 1\frac{1}{3} =$ _____

10. $9\frac{11}{12} - 4\frac{1}{12} =$ _____

11. $3\frac{60}{100} - 1\frac{20}{100} =$ _____

Problem Solving

Solve. Write in simplest form.

12. Isabella has $2\frac{1}{4}$ oranges. Colleen has $3\frac{1}{4}$ oranges. How many oranges do they have all together?

Brain Builders

13. Liam had $5\frac{7}{8}$ cups of flour. He used $2\frac{3}{8}$ cups of flour to make bread. He used $1\frac{5}{8}$ cups of flour to make muffins. How much flour does Liam have left?

14. Audrey ran $2\frac{1}{6}$ miles yesterday. Today, she ran $1\frac{3}{6}$ miles. Tomorrow, she plans on running $1\frac{4}{6}$ miles less than yesterday and today combined. How many miles will Audrey run tomorrow?

15. **Test Practice** Isaac's paper airplane flew $5\frac{2}{12}$ feet. Lillian's paper airplane flew $5\frac{5}{12}$ feet. How much farther did Lillian's paper airplane fly?

    Ⓐ $\frac{1}{4}$ foot

    Ⓑ $\frac{3}{5}$ foot

    Ⓒ $5\frac{1}{12}$ feet

    Ⓓ $5\frac{7}{12}$ feet

Name _____

Lesson 9
Multiply Fractions by Whole Numbers

ESSENTIAL QUESTION

How can I use operations to model real-world fractions?

You can use models and equations to multiply a fraction by a whole number.

Math in My World

Example 1

Each card on a trivia game has 6 questions. Each question represents $\frac{1}{6}$ of the questions on the card. Caleb correctly answered 4 of the questions. What fraction of the questions on a card did he answer correctly?

Find $4 \times \frac{1}{6}$.

One Way

Use repeated addition.

Use repeated addition to write an equation.

$$4 \times \frac{1}{6} = \frac{1}{6} + \frac{1}{6} + \frac{1}{6} + \frac{1}{6}$$

$$= \frac{4}{6} \qquad \text{Add like fractions.}$$

$$= \frac{\Box}{\Box} \qquad \text{Simplify.}$$

So, $4 \times \frac{1}{6} = \frac{\Box}{\Box}$.

Another Way

Use models.

The number line shows the first four multiples of $\frac{1}{6}$.

Check Use fraction tiles. $4 \times \frac{1}{6} = \frac{4}{6}$ or $\frac{\Box}{\Box}$

Online Content at **connectED.mcgraw-hill.com**

You can use equations and properties to multiply a fraction by a whole number.

Example 2

Find $5 \times \frac{3}{10}$. Identify the two whole numbers between which the product lies.

Think of $\frac{3}{10}$ as a multiple of $\frac{1}{10}$.

$5 \times \frac{3}{10} = 5 \times \left(3 \times \frac{1}{10}\right)$ $\quad\quad \frac{3}{10} = 3 \times \frac{1}{10}$

$\quad\quad\quad = (5 \times 3) \times \frac{1}{10}$ $\quad\quad$ Associative Property

$\quad\quad\quad = 15 \times \frac{1}{10}$ $\quad\quad\quad\quad$ Multiply. $5 \times 3 = 15$

$\quad\quad\quad = \frac{15}{10}$ $\quad\quad\quad\quad\quad\quad$ 15 groups of $\frac{1}{10} = \frac{15}{10}$

$\quad\quad\quad = 1\frac{5}{10}$ $\quad\quad\quad\quad\quad\quad$ 15 divided into groups of 10 $= 1\frac{5}{10}$

$\quad\quad\quad = 1\frac{1}{2}$ $\quad\quad\quad\quad\quad\quad$ Simplify. $\frac{5}{10} = \frac{1}{2}$

So, $5 \times \frac{3}{10} =$ _____ .

The product lies between the whole numbers 1 and 2.

Look at the product before it was simplified.

$5 \times \frac{3}{10} = \frac{15}{10}$ ← The numerator of the product is the same as the product of the whole number and the _____ of the fraction. $5 \times 3 = 15$

The denominator of the product is the same as the _____ of the fraction.

Guided Practice

Multiply.

1. $5 \times \frac{1}{8} =$ _____

2. $4 \times \frac{2}{3} =$ _____

Does $3 \times \frac{7}{8} = 3\frac{7}{8}$? Explain.

614 Chapter 9 Operations with Fractions

Review

Chapter 9

Operations with Fractions

Vocabulary Check

1. Use the numbers on the number cubes to create each type of fraction listed below. You can use the numbers more than once.

like fractions

_____

mixed number

_____

simplest form

_____

Concept Check

Find each sum. Write in simplest form.

2. $\frac{1}{6} + \frac{4}{6} =$ _____

3. $3\frac{3}{5} + 2\frac{1}{5} =$ _____

4. $\frac{2}{8} + \frac{4}{8} =$ _____

My Chapter Review 619

Find each difference. Write in simplest form.

5. $\dfrac{3}{4} - \dfrac{2}{4} =$ _____

6. $5\dfrac{2}{3} - 2\dfrac{1}{3} =$ _____

7. $\dfrac{8}{10} - \dfrac{3}{10} =$ _____

Algebra Use an equation to write each fraction or product as a multiple of a unit fraction.

8. $\dfrac{3}{5}$ _____

9. $\dfrac{9}{10}$ _____

10. $\dfrac{8}{12}$ _____

11. $\dfrac{7}{8}$ _____

12. $4 \times \dfrac{3}{5}$ _____

13. $2 \times \dfrac{9}{10}$ _____

14. $3 \times \dfrac{8}{12}$ _____

15. $6 \times \dfrac{7}{8}$ _____

Multiply. Write in simplest form. Identify the two whole numbers between which the product lies.

16. $6 \times \dfrac{1}{4} =$ _____

17. $24 \times \dfrac{1}{5} =$ _____

18. $7 \times \dfrac{3}{5} =$ _____

19. $3 \times \dfrac{4}{6} =$ _____

620 **Chapter 9** Operations with Fractions

Name _____

Problem Solving

20. Kylie and her two sisters each have $\frac{1}{4}$ cup of chopped pineapple. How much pineapple do they have in all?

_____

21. Joshua and his four friends each have $\frac{1}{2}$ package of crackers. How many packages of crackers do they have in all? Explain.

_____

_____

Brain Builders

22. Write a real-world problem to add like fractions. Then solve the problem.

_____

_____

_____

23. Test Practice There are 25 students in Miss Klempa's class. Suppose each student has $\frac{3}{10}$ of a box of pencils. How many boxes of pencils do they have in all?

Ⓐ 6 boxes   Ⓒ 9 boxes

Ⓑ $7\frac{1}{2}$ boxes   Ⓓ $10\frac{1}{2}$ boxes

Reflect

Chapter 9
Answering the
ESSENTIAL QUESTION

Use what you learned about using operations with fractions to complete the graphic organizer.

ESSENTIAL QUESTION

How can I use operations to model real-world fractions?

Real-World Example → **Using Operations**

Real-World Example → **Using Operations**

Real-World Example → **Using Operations**

Reflect on the ESSENTIAL QUESTION Write your answer below.

_____
_____

622 Chapter 9 Operations with Fractions

Name _____ Date _____

Score _____

Performance Task

Clara the Cleaner

Clara is an amazing cleaner. All her success comes from the cleaning products she uses. They are all made with her secret ingredients.

Show all your work to receive full credit.

Part A

Clara needed to clean a small office building. She added $\frac{3}{4}$ cup of Secret Ingredient #1 to each of 9 buckets of soapy water. How much Secret Ingredient #1 did she use in all?

Complete the model and write and solve an equation. Show your work.

Part B

Clara found a box of Secret Ingredient #2 that was $\frac{5}{6}$ full, and another box that was $\frac{4}{6}$ full. She wants to put $\frac{2}{6}$ box in each of 4 buckets of glass cleaner. Does she have enough of Secret Ingredient #2 to fill all the buckets? Explain.

Part C

Clara needs to add $\frac{3}{5}$ teaspoon of Secret Ingredient #3 to each small bucket to polish the furniture. She has 8 buckets ready to go, but she only has 4 teaspoons of the ingredient. Does she have enough for 8 buckets? Solve the problem 2 different ways. Tell how much she has left over or how much more she needs.

Part D

Clara adds $\frac{3}{8}$ cup of Secret Ingredient #4 to each of 4 buckets to shine the elevator doors. She got a new job to clean another larger building and now she needs enough Secret Ingredient #4 to make an additional 5 buckets. Write and solve an equation to find how much of the ingredient she needs in all.

Name
....

MY Chapter Project

Bake Sale Equivalents

1. With your partner, list 5 baked goods you would like to sell at a bake sale. Include a price for each item. Write the prices in decimal form.

Item	Price

2. List the foods with their prices in order from least to greatest:

_____

_____

_____

_____

3. Make a poster that includes your food names and prices listed in order from least to greatest. Include a picture of each food.

624 Chapter 10 Fractions and Decimals

Name _____

Am I Ready?

Write a fraction to describe the part that is green.

1.

2.

3.

_____   _____   _____

Write each as a fraction.

4. four tenths

5. eight tenths

6. twenty hundredths

_____   _____   _____

7. On Tuesday, seven tenths of an inch of rain fell. Write the amount of rain as a fraction.

_____

Algebra Find each unknown.

8.

$\frac{1}{5} = \frac{\square}{10}$

9.

$\frac{4}{5} = \frac{\square}{10}$

Shade the boxes to show the problems you answered correctly.

How Did I Do? 1 2 3 4 5 6 7 8 9

Online Content at connectED.mcgraw-hill.com

Name

MY Math Words

Review Vocabulary

equivalent   fraction   place value

Making Connections

Use the review vocabulary to make equivalent fractions. Find the unknown in the fraction or shade to complete the area model.

Tenths	Hundredths
$\frac{3}{10}$	= $\frac{\boxed{}}{100}$
$\frac{8}{\boxed{}}$	= $\frac{80}{100}$

626 Chapter 10 Fractions and Decimals

MY Vocabulary Cards

Lesson 10-1

decimal

0.75 0.75

Lesson 10-1

hundredth

0.73 0.73

Lesson 10-1

tenth

0.8 0.8

Ideas for Use

- Draw or write examples for each card. Be sure your examples are different from what is shown on each card.
- Use a blank card to write a word from a previous chapter that you would like to review.
- Use a blank card to write this chapter's essential question. Use the back of the card to write or draw examples that help you answer the question.

One of one hundred equal parts.

Write *thirty-one hundredths* as a decimal.

_____

A number that uses place value and a decimal point to show part of a whole.

Explain how decimals and fractions are alike.

_____
_____

One of ten equal parts.

Write *six tenths* as a decimal.

_____

MY Foldable

FOLDABLES Follow the steps on the back to make your Foldable.

Name _____

Lesson 1
Hands On
Place Value Through Tenths and Hundredths

ESSENTIAL QUESTION
How are fractions and decimals related?

A **decimal** is a number that uses place value and a decimal point to show part of a whole.

Build It

Kendra has 1 dollar, 3 dimes, and 5 pennies. Model this amount using a place-value chart.

You know that one dollar is equal to _____ cents.

One dollar is one whole. So, write 1 in the ones place.

One dime is one of 10 equal parts of a dollar. One of 10 equal parts is one **tenth**.

So, one dime is *one tenth*, or $\frac{1}{10}$, of a dollar.

Kendra has _____ dimes. Write 3 in the tenths place.

One penny is one of 100 equal parts of a dollar. One of 100 equal parts is one **hundredth**.

So, one penny is *one hundredth*, or $\frac{1}{100}$, of a dollar.

Kendra has _____ pennies. Write 5 in the hundredths place.

The decimal that represents the amount of money Kendra has, in dollars, is $1.35.

1 dollar | 5 pennies
3 dimes

Online Content at connectED.mcgraw-hill.com

Lesson 1 631

You can use models to represent decimals.

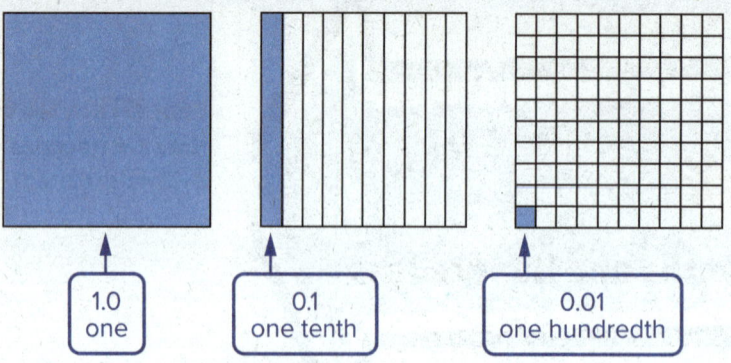

Try It

Complete the place-value chart that represents the fraction of the grid that is shaded at the right.

There are _____ squares shaded out of a total of

_____ squares.

In words, this is *forty-two hundredths*.

This is the same as 4 tenths and 2 hundredths.

So, write 4 tenths and 2 hundredths in the place-value chart.

4 tenths   2 hundredths

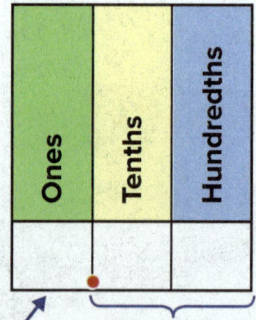

There are 0 whole grids shaded.   4 tenths and 2 hundredths is 42 hundredths.

Talk About It

1. **Processes & Practices 2** **Use Number Sense** Paulo has 6 dimes. Marc has 6 pennies. How many times greater is the value of 6 dimes than 6 pennies? Explain.

_____

_____

632 Chapter 10 Fractions and Decimals

Practice It

Complete the place-value chart that represents each set of dollar bills and coins.

2. 0 dollars, 4 dimes, 8 pennies

3. 2 dollars, 9 dimes, 1 penny

Write the decimal represented by each model.

4.

5.

_____

Write each decimal.

6. two tenths

_____

7. twenty-five hundredths

_____

8. forty-nine hundredths

_____

9. seventy-three hundredths

_____

Shade each model to represent each decimal.

10. 0.7

11. 0.51

12. Circle the greater decimal in Exercises 10 and 11.

Lesson 1 Hands On: Place Value Through Tenths and Hundredths 633

Apply It

13. A flower has ten petals. During a storm, five of the petals fall off. Write a decimal to show what part of the petals is left.

5 petals fell off

_____

14. A movie theater has 100 seats. Seventy-eight of the seats are full. What decimal shows the part of the seats that are full?

78 seats are full

_____

15. Processes & Practices 4 Model Math Write a real-world problem that results in an answer of 0.58.

_____
_____

Write About It

16. How can I relate decimals to a place-value chart?

_____
_____

634 Chapter 10 Fractions and Decimals

Name _____

Lesson 1

Hands On: Place Value Through Tenths and Hundredths

Homework Helper   Need help? connectED.mcgraw-hill.com

Complete the place-value chart to represent the fraction of the grid that is shaded.

There are 46 squares shaded out of a total of 100 squares.

In words, this is *forty-six hundredths*.

This is the same as 4 tenths and 6 hundredths.

Write 4 tenths and 6 hundredths in the place-value chart below.

4 tenths   6 hundredths

There are 0 whole grids shaded.

4 tenths and 6 hundredths is 46 hundredths.

Practice

Write the decimal represented by each model.

1.

2.

Lesson 1 My Homework 635

Problem Solving

3. Kristy has 100 buttons in her button collection. She has 24 red buttons. Write a decimal to show the part of her collection of buttons that is red.

   _____

4. Harry and Dario went to the county fair. There were ten rides, but they only had time to ride six. What decimal shows the part of the rides that they rode?

   _____

5. There are ten swings on the playground. Eight children are swinging on the swings. Write a decimal to show the part of the swings that are being used.

   _____

6. There are 100 students in the fourth grade. Five of these students are absent today. Write a decimal to show the part of the students that are absent.

   _____

Vocabulary Check

Draw a line to match each word to its definition.

7. decimal

8. tenth

9. hundredth

- one part of ten equal parts
- a number that uses place value and a decimal point to show part of a whole
- one part of one hundred equal parts

Name _____

Lesson 2
Tenths

ESSENTIAL QUESTION
How are fractions and decimals related?

Math in My World

Example 1

Lorena and Rebecca went to 10 horseback riding lessons. Three of the lessons were outside the barn. What decimal represents the part of the horseback riding lessons that were spent outside?

There were 10 horseback riding lessons. Three of the 10 lessons were spent outside. Three out of ten equal parts means *three tenths*.

 Use a model.
Three tenths of the model is shaded.

 Use a place-value chart.
Write 3 tenths in the place-value chart.

Ones	Tenths	Hundredths
	.	

Read this decimal as *three tenths*.

There are 0 whole grids shaded. There are 3 tenths shaded.

So, _____ of the horseback riding lessons were outside.

Online Content at connectED.mcgraw-hill.com

Lesson 2   637

Example 2

A campground has 10 campsites. Eight of the campsites have campers. The rest are empty. What part of the campground is empty? Write as a decimal. Use a number line.

If _____ out of ten campsites have campers, then _____ out of ten campsites do *not* have campers.

The number line is divided from 0 to 1 in 10 equal parts.

Two out of ten equal parts is *two tenths*, or 0.2.

So, _____ of the campground is empty.

Example 3

Use words to describe 0.7.

The 7 is in the tenths place. So, 0.7 is *seven tenths*. *Seven tenths* means seven out of ten equal parts.

So, 0.7 can be described as _____ out of _____.

Guided Practice

Shade the model. Then write the decimal.

1. four out of ten

2. eight tenths

Talk MATH

Explain why dimes could be used to model tenths.

638 Chapter 10 Fractions and Decimals

Name

Independent Practice

Shade the model. Then write the decimal.

3. one out of ten

 _____

4. five out of ten

 _____

5. two tenths

 _____

6. seven tenths

 _____

Write each of the following as a decimal. Then graph each point on the same number line.

7. six out of ten equal parts

8. nine out of ten equal parts

_____   _____

Use words to describe each decimal.

9. 0.2

10. 0.3

_____   _____

11. 0.8

12. 0.9

_____   _____

Lesson 2 Tenths 639

Problem Solving

13. Processes &Practices 1 **Make Sense of Problems** Cody wanted to play fetch with his dog. His dog had ten chew toys. Cody took six toys outside for his dog. Write a decimal to show what part of the toys is still inside.

14. Emilio had a math quiz today. He answered nine of the ten questions correctly. Write a decimal to show what part of the quiz Emilio answered correctly.

Brain Builders

15. Processes &Practices 4 **Model Math** Jasmine had ten pennies in her pocket. Some fell through a hole in her pocket. Below are the pennies she still has in her pocket. Draw a model to represent the part of the pennies that are still in her pocket. Use the model to help write a decimal to show what part of the pennies she lost.

16. Processes &Practices 2 **Use Number Sense** Write a decimal whose value is greater than five tenths. Explain how the value of your decimal compares to four tenths.

17. ❓ Building on the Essential Question How can I use number lines and decimal grids to model tenths?

640 Chapter 10 Fractions and Decimals

Name _____

MY Homework

Lesson 2

Tenths

Homework Helper

Need help? connectED.mcgraw-hill.com

Model and write *six out of ten* as a decimal.
Six tenths of the model is shaded.

There are 0 whole grids shaded.
6 tenths are shaded.

So, six out of ten is 0.6.

Practice

Model and write each decimal.

1. two out of ten

2. four out of ten

Use words to describe each decimal.

3. 0.3

4. 0.8

Lesson 2 My Homework 641

Write each of the following as a decimal. Then graph each point on the same number line.

5. one out of ten equal parts

6. seven out of ten equal parts

_____   _____

0 0.1 0.2 0.3 0.4 0.5 0.6 0.7 0.8 0.9 1

Problem Solving

7. Traci is playing a game at the circus. She has ten chances to throw a dart at a balloon. She hits the balloon 8 out of ten times. Write a decimal to show how many times Traci hit the balloons.

_____

8. Jackie took a science quiz. She only answered one out of ten problems incorrectly. Write a decimal to show the part of the quiz that Jackie answered correctly.

_____

Brain Builders

9. **Processes & Practices** 4 **Model Math** Carter's mom bought ten fruits at the grocery store. Seven of the fruits are bananas. Write a decimal to show the part of the fruits that are *not* bananas.

_____

10. A parking lot has ten parking spaces. There are the same number of cars parked in the parking lot as there are empty parking spaces. Write a decimal to show the part of the parking lot that has empty parking spaces.

11. **Test Practice** Which choice represents the part of the model that is shaded?

Ⓐ four one-hundredths
Ⓑ 1 out of 4
Ⓒ 0.4
Ⓓ 4.0

Name ....

Lesson 3
Hundredths

ESSENTIAL QUESTION
How are fractions and decimals related?

Math in My World

Example 1

Drew has a 100-piece puzzle. There are four corner pieces in the puzzle. What decimal represents the part of the puzzle that is corner pieces?

Four of the 100 puzzle pieces are corner pieces.

 Use a model.

Shade 4 of the 100 parts.

 Use a place-value chart.

Write 4 hundredths in the place-value chart.

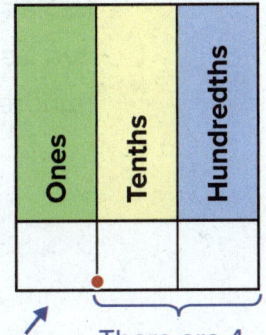

Read this decimal as *four hundredths*.

There are 0 whole grids shaded.
There are 4 hundredths shaded.

So, _____ of the puzzle pieces are corner pieces.

Online Content at connectED.mcgraw-hill.com

Lesson 3 643

Example 2

It rained 13 out of the past 100 days. What part of the days did it rain? Write as a decimal.

 Use a model.

The model shows 13 parts of 100 that are shaded.

The model shows _____ out of 100.

Thirteen out of 100 equal parts is *thirteen hundredths*, or _____.

 Use a place-value chart.

Write 13 hundredths in the place-value chart.

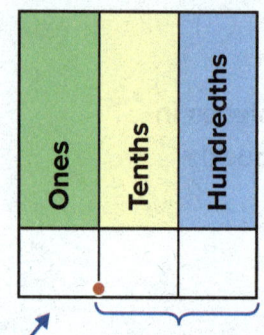

There are 0 whole grids shaded.

13 hundredths is 1 tenth and 3 hundredths.

Read this decimal as *thirteen hundredths*.

So, it rained _____ of the days.

Guided Practice

Write each decimal.

1.

2.

Talk MATH

Explain why pennies can be used to model hundredths.

644 Chapter 10 Fractions and Decimals

Name ....

Independent Practice

Write each decimal.

3.
_____

4.
_____

5.
_____

6.
_____

7.
_____

8.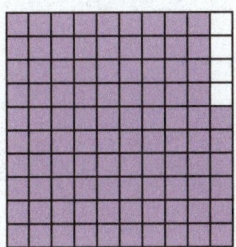
_____

Write a decimal for each part of a dollar shown.

9.
_____

10.
_____

11.
_____

12.
_____

Lesson 3 Hundredths 645

Problem Solving

13. Emma earned eighty-nine out of 100 points on a math test. What decimal shows the part of the points that Emma earned?

_____

14. A department store had one hundred suitcases. Six of the suitcases were sold. What decimal represents the part of the suitcases that were left?

_____

15. Bena received the coins shown as change after buying pencils at the school store. What part of a dollar are the coins?

_____

Brain Builders

16. **Processes &Practices** **Reason** How are forty hundredths and four tenths related? Explain.

_____

_____

17. **Building on the Essential Question** How can I use decimal grids to model hundredths? Explain how a number line could be used to model hundredths.

_____

_____

_____

_____

646 **Chapter 10** Fractions and Decimals

Name ....

MY Homework

Lesson 3

Hundredths

Homework Helper

Need help? connectED.mcgraw-hill.com

On a video game, Sandra scored 72 points out of a possible 100 points. What part of the possible points did she score? Write as a decimal.

1 Use a model.
The model shows 72 shaded parts out of 100.
This is *seventy-two hundredths*, or 0.72.

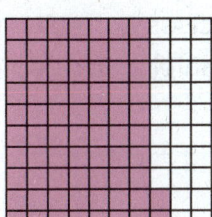

2 Use a place-value chart.
Write 72 hundredths in the place-value chart.

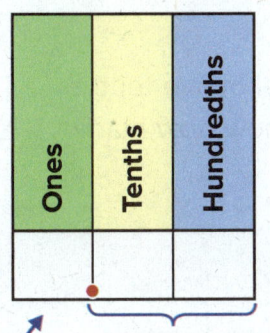

There are 0 whole grids shaded.

72 hundredths is 7 tenths and 2 hundredths.

Read this decimal as *seventy-two hundredths*.

So, Sandra scored 0.72 of the possible points.

Practice

Write a decimal for each part of a dollar shown.

1.

_____

2.

_____

Lesson 3 My Homework 647

Problem Solving

3. Dexter bought an apple. He paid using the coins shown below.

Write a decimal to show the part of a dollar that he paid.

_____

4. Daniel had a social studies quiz. He answered eighty-seven out of 100 exercises correctly. Write a decimal to show the part of the quiz that Daniel answered correctly.

_____

Brain Builders

5. **Processes & Practices** **4** **Model Math** Claire drew a model of five tenths. Explain how Claire could change her model to show how many hundredths are in five tenths.

_____
_____

6. **Processes & Practices** **6** **Be Precise** Ginger and Caleb each have 0.77 of a dollar. Ginger does not have any quarters. Caleb does not have any dimes. Explain how many of each type of coin each of them might have.

_____
_____

7. **Test Practice** Which decimal represents the part of the model that is shaded?

Ⓐ 0.57 Ⓒ 5.70
Ⓑ 0.75 Ⓓ 57.0

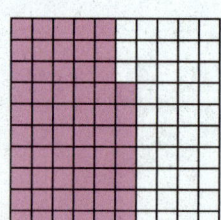

648 **Need more practice?** Download Extra Practice at connectED.mcgraw-hill.com

Check My Progress

Vocabulary Check

Write *true* or *false*.

1. A **decimal** is a number that uses place value and a decimal point to show part of a whole. _____

2. In the number 0.36, the 3 is in the **tenths** place. _____

3. In the number 0.58, the 5 is in the **hundredths** place. _____

Concept Check

Write each decimal.

4. four tenths _____

5. thirty-two hundredths _____

Shade the model. Then write the decimal.

6. three out of ten

_____

7. eight out of ten

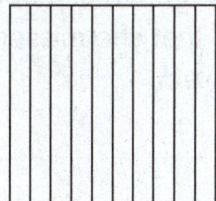

_____

Write each decimal.

8. _____

9. 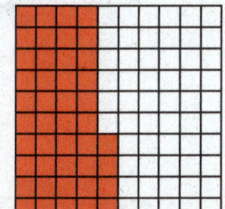 _____

Check My Progress 649

Problem Solving

10. The distance between the grocery store and the post office is thirty-four hundredths of a mile. Write this distance as a decimal.

11. There are ten fish in an aquarium. Four of them are blue. Write a decimal to show the part of the fish that are blue.

Brain Builders

12. William found the coins shown in his pocket. Use a grid to show the part of a dollar represented by the coins. Use a different color to shade each coin. Write a decimal to show the part of a dollar represented by each individual coin and the total of all of the coins.

13. There are ten questions on a test. Sophie answered 8 of them correctly. Sophie says that she missed 0.3 out of all of the questions. Is Sophie correct? Explain.

14. **Test Practice** Which of the following is represented by the number line?

    Ⓐ 4.0
    Ⓑ four tenths
    Ⓒ 4 out of 5
    Ⓓ four hundredths

650 Chapter 10 Fractions and Decimals

Name _____

Lesson 4
Hands On
Model Decimals and Fractions

ESSENTIAL QUESTION
How are fractions and decimals related?

You can write a decimal as a fraction. Both a decimal and a fraction show part of a whole.

Draw It

Write 0.4 as a fraction with a denominator of 10.

 Write the decimal in the place-value chart.

0.4 means *four tenths.*

 Shade the grid to represent four tenths.

Shade four of the ten equal parts.

 Write a fraction.

← number of shaded parts
← total number of parts

Think of 0.4 as *four tenths* or $\frac{4}{10}$.

Check

The number line shows that 0.4 is 4 out of 10 or $\frac{4}{10}$.

Online Content at connectED.mcgraw-hill.com

Try It

Write 0.45 as a fraction with a denominator of 100.

Use a number line.

So, 0.45 can be written as .

Talk About It

1. Shade the models below to show how 0.7 and 0.70 are equivalent.

2. Write two fractions that represent the models in Exercise 1.

_____

3. **Processes & Practices 3** **Draw a Conclusion** Explain how you could write $\frac{9}{10}$ as a fraction with a denominator of 100 without using models.

_____

_____

4. Explain how you could write 0.78 as a fraction with a denominator of 100 without using models.

_____

Name

Practice It

Write each decimal as a fraction with a denominator of 10. Shade the grid.

5. 0.2 = _____

6. 0.7 = _____

7. 0.3 = _____

8. 0.5 = _____

9. 0.6 = _____

10. 0.1 = _____

Write each decimal as a fraction with a denominator of 100. Shade the grid.

11. 0.39 = _____

12. 0.71 = _____

13. 0.12 = _____

14. 0.09 = _____

15. 0.23 = _____

16. 0.02 = _____

Lesson 4 Hands On: Model Decimals and Fractions

Apply It

17. There are 10 cars on the racetrack. If 3 of the cars are red, what fraction of the cars are not red? Write as a fraction with a denominator of 10.

_____

18. On Monday, it snowed 0.6 of an inch. Write 0.6 as a fraction with a denominator of 10.

_____

19. There are 100 students at the gymnastics competition. 57 hundredths, or 0.57 of the students, are nine years old. Write 0.57 as a fraction with a denominator of 100.

_____

20. The fourth grade class is donating 100 clothing items to charity. 63 hundredths of the items are sweaters and 0.20 of the items are jeans. What part of the clothing items are neither sweaters nor jeans? Write the number as a fraction with a denominator of 100.

_____

21. Processes &Practices 6 **Explain to a Friend** Explain to a classmate why 0.8 and 0.80 name the same number.

_____
_____
_____

Write About It

22. How can I use models to relate fractions and decimals?

_____
_____
_____

654 Chapter 10 Fractions and Decimals

Name _____

MY Homework

Lesson 4

Hands On:
Model Decimals
and Fractions

Homework Helper

Need help? connectED.mcgraw-hill.com

Write 0.7 as a fraction with a denominator of 10.

1 Write the decimal in the place-value chart.

0.7 means *seven tenths*.

Ones	Tenths	Hundredths
0	7	

2 Use a grid to represent seven tenths.

The model shows seven of the ten equal parts are shaded.

3 Write a fraction.

$\frac{7}{10}$ ← number of shaded parts
         ← total number of parts

Think of *seven tenths* as $\frac{7}{10}$.

So, $0.7 = \frac{7}{10}$.

Check

The number line shows that 0.7 is 7 out of 10 or $\frac{7}{10}$.

0   0.1 0.2 0.3 0.4 0.5 0.6 0.7 0.8 0.9 1

7 parts out of 10 equal parts

Lesson 4 My Homework 655

Practice

Write each decimal as a fraction with a denominator of 10. Shade the grid.

1. 0.1 = _____

2. 0.3 = _____

3. 0.9 = _____

Write each decimal as a fraction with a denominator of 100. Shade the grid.

4. 0.17 = _____

5. 0.24 = _____

6. 0.88 = _____

Problem Solving

 Josiah's family has 10 pets. For Exercises 7–9, refer to the table that shows how many of each type of pet Josiah's family has.

Type of Pet	Part
Cat	0.2
Dog	0.1
Rabbit	0.2
Fish	0.4
Reptile	0.1

7. Write the part of the family's pets that are cats as a fraction with a denominator of 10.

_____

8. Write the part of the family's pets that are fish as a fraction with a denominator of 10.

_____

9. Write the part of the family's pets that are fish as a fraction with a denominator of 100.

_____

656 Chapter 10 Fractions and Decimals

Name _____

Lesson 5
Decimals and Fractions

ESSENTIAL QUESTION
How are fractions and decimals related?

Decimals and fractions can show equivalent amounts. You can write equivalent fractions and you can also write fractions as decimals.

Math in My World

Example 1

Kara travels $\frac{7}{10}$ mile from her bus stop to the next stop. Write $\frac{7}{10}$ as a fraction with a denominator of 100. Then write the fraction as a decimal.

Use a model to show $\frac{7}{10}$.

1 Write $\frac{7}{10}$ as a fraction with a denominator of 100.

Shade seven tenths of the tenths grid.

On the hundredths grid, shade squares so that the same fraction of the two grids is shaded.

How many squares did you shade on the hundredths grid? _____

The decimal models show that $\frac{7}{10}$ is equivalent to $\frac{70}{100}$.

2 Write the fraction as a decimal.

Think of $\frac{7}{10}$ as *seven tenths*. So, $\frac{7}{10} = 0.7$.

So, $\frac{7}{10}$ can be written as $\frac{70}{100}$ and 0.7.

Helpful Hint
Notice that seventy hundredths, or 0.70, is the same as seven tenths, or 0.7.

Online Content at connectED.mcgraw-hill.com

Lesson 5 657

Example 2

Write $\frac{9}{10}$ as an equivalent fraction with a denominator of 100 and as a decimal.

Use a model to show $\frac{9}{10}$.

 Write $\frac{9}{10}$ as a fraction with a denominator of 100.

Shade nine tenths of the tenths grid.

On the hundredths grid, shade squares so that the same fraction on the two grids is shaded.

How many squares did you shade on the

hundredths grid? _____

The decimal models show that $\frac{9}{10}$ is equivalent to $\frac{90}{100}$.

 Write the fraction as a decimal.

Think of $\frac{9}{10}$ as *nine tenths*. So, $\frac{9}{10} = 0.9$.

So, $\frac{9}{10}$ can be written as $\frac{90}{100}$ and 0.9.

Notice that ninety hundredths, or 0.90, is the same as *nine tenths*, or 0.9.

Guided Practice

Write each fraction as an equivalent fraction with a denominator of 100. Then write the fraction as a decimal.

1. $\frac{1}{10}$

   Fraction:

   Decimal: _____

2. $\frac{5}{10}$

   Fraction:

   Decimal: _____

Talk MATH

Is one tenth greater than or less than one hundredth? Explain.

658 **Chapter 10** Fractions and Decimals

Name ....

Independent Practice

Write each fraction as an equivalent fraction with a denominator of 100. Shade the grids to show that the fractions are equivalent.

3. $\frac{1}{10} =$ _____

4. $\frac{8}{10} =$ _____

5. $\frac{5}{10} =$ _____

6. Write the fraction in Exercise 3 as a decimal. _____

7. Write the fraction in Exercise 4 as a decimal. _____

8. Write the fraction in Exercise 5 as a decimal. _____

Circle the equivalent ways to represent each number.

9. $\frac{6}{10}$ six tenths   six hundredths $\frac{6}{100}$ 0.6   0.06

10. $\frac{30}{100}$ $\frac{3}{10}$ thirty hundredths $\frac{3}{100}$ 0.03   0.3

11. 0.2   two tenths   twenty hundredths $\frac{2}{100}$ 0.20   0.02

12. 0.7 $\frac{70}{100}$ seven hundredths $\frac{7}{100}$ $\frac{7}{10}$ 0.70

Lesson 5 Decimals and Fractions

Problem Solving

13. Wesley learned that $\frac{3}{10}$ of the students in his class are left handed. Write $\frac{3}{10}$ as a fraction with a denominator of 100 and as a decimal.

14. Processes &Practices 2 Use Algebra Tyrone is completing ■ $= \frac{5}{100}$. What is the unknown decimal equivalent?

Brain Builders

15. Of Marissa's doll collection, $\frac{4}{10}$ are dolls with brown hair. Write the number of Marissa's dolls that do *not* have brown hair as a fraction with a denominator of 10 and as a fraction with a denominator of 100. Then write the fraction as a decimal.

16. Processes &Practices 6 Be Precise Explain the similarities and differences between five tenths and five hundredths.

17. Building on the Essential Question Why can I identify the shaded part of a hundredths grid as a fraction and a decimal?

660 Chapter 10 Fractions and Decimals

Name _____

MY Homework

Lesson 5
Decimals and Fractions

Homework Helper

Need help? connectED.mcgraw-hill.com

Write $\frac{5}{10}$ as an equivalent fraction with a denominator of 100 and as a decimal.

Use a model to show $\frac{5}{10}$.

1 Write $\frac{5}{10}$ as a fraction with a denominator of 100.

Five tenths of the tenths grid are shaded.

On the hundredths grid, squares were shaded so that the same fraction on the two grids is shaded.

Fifty squares on the hundredths grid are shaded.

The decimal models show that $\frac{5}{10}$ is equivalent to $\frac{50}{100}$.

2 Write the fraction as a decimal.

Think of $\frac{5}{10}$ as *five tenths*. So, $\frac{5}{10}$ = 0.5.

So, $\frac{5}{10}$ can be written as $\frac{50}{100}$ and 0.5.

Practice

1. Write $\frac{4}{10}$ as an equivalent fraction with a denominator of 100. Shade the grids to show that the fractions are equivalent. Then write the fraction as a decimal.

Fraction: ⬜/⬜

Decimal: _____

Lesson 5 My Homework 661

Problem Solving

2. Of Cadence's paintbrushes, $\frac{2}{10}$ were used to paint a picture. Write $\frac{2}{10}$ as a fraction with a denominator of 100. Then write the fraction as a decimal.

_____

3. Of the trees in a park, $\frac{6}{10}$ have red leaves. Write $\frac{6}{10}$ as a fraction with a denominator of 100. Then write the fraction as a decimal.

_____

Brain Builders

4. Of the ten menu items at a deli, 7 are sandwiches with meat, 2 are sandwiches without meat, and 1 is a dessert. Write the fraction of menu options that are sandwiches as a fraction with a denominator of 100. Then write the fraction as a decimal.

_____

5. **Processes &Practices 3** **Find the Error** Marjorie wrote the fraction $\frac{7}{10}$ as a fraction with a denominator of 100. She then wrote the fraction as a decimal. Find and correct her mistake. Explain. Include a model to support your answer.

$$\frac{7}{10} = \frac{7}{100}$$
$$= 0.07$$

_____
_____
_____

6. **Test Practice** Which fraction is equivalent to the fraction that represents 8 correct answers out of 10?

Ⓐ $\frac{8}{100}$ Ⓑ $\frac{80}{100}$ Ⓒ $\frac{8}{1}$ Ⓓ $\frac{80}{10}$

Name _____

Lesson 6
Use Place Value and Models to Add

ESSENTIAL QUESTION
How are fractions and decimals related?

Math in My World

Example 1

Denny walked $\frac{3}{10}$ mile to the post office. Then he walked $\frac{5}{100}$ mile to the grocery store. How far did he walk in all? Write the answer as a fraction with a denominator of 100 and as a decimal.

Use a model to show $\frac{3}{10} + \frac{5}{100}$.

1 Write $\frac{3}{10}$ as a fraction with a denominator of 100.

The decimal models show that $\frac{3}{10} = \frac{30}{100}$.

2 Add like fractions.

$\frac{30}{100} + \frac{5}{100} = \frac{30 + 5}{100}$

$= \frac{\square}{100}$

3 Write the sum as a decimal.

Think of $\frac{35}{100}$ as *thirty-five hundredths*. So, $\frac{35}{100} = 0.35$.

So, $\frac{3}{10} + \frac{5}{100} = \frac{35}{100}$, or _____.

Denny walked $\frac{35}{100}$, or 0.35, mile in all.

Online Content at connectED.mcgraw-hill.com

Lesson 6 663

Example 2

Find $\frac{4}{10} + \frac{22}{100}$. Write the sum as a fraction with a denominator of 100 and as a decimal.

1 Write $\frac{4}{10}$ as a fraction with a denominator of 100.

The decimal models show that $\frac{4}{10} = \frac{40}{100}$.

2 Add like fractions.

$\frac{40}{100} + \frac{22}{100} = \frac{40 + 22}{100}$

$= \frac{\boxed{}}{100}$

3 Write the sum as a decimal.

Think of $\frac{62}{100}$ as *sixty-two hundredths*. So, $\frac{62}{100} = 0.62$.

So, $\frac{4}{10} + \frac{22}{100} = \frac{62}{100}$ or _____.

Check The models show that $\frac{4}{10} + \frac{22}{100} = \frac{62}{100}$, or 0.62.

 + =

Guided Practice

Add. Write each sum as a fraction with a denominator of 100 and as a decimal.

1. $\frac{5}{10} + \frac{1}{100} =$ _____

2. $\frac{7}{10} + \frac{13}{100} =$ _____

In Example 2, why was $\frac{4}{10}$ written as $\frac{40}{100}$?

664 Chapter 10 Fractions and Decimals

Name ....

Independent Practice

Shade the models to find each sum. Write the sum as a fraction with a denominator of 100.

3. $\frac{2}{10} + \frac{37}{100} =$ _____

4. $\frac{7}{10} + \frac{11}{100} =$ _____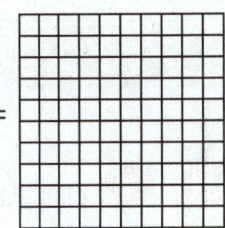

Add. Write each sum as a fraction with a denominator of 100 and as a decimal.

5. $\frac{6}{10} + \frac{24}{100} =$ _____

6. $\frac{5}{10} + \frac{21}{100} =$ _____

7. $\frac{3}{10} + \frac{65}{100} =$ _____

8. $\frac{1}{10} + \frac{52}{100} =$ _____

9. $\frac{8}{10} + \frac{17}{100} =$ _____

10. $\frac{2}{10} + \frac{42}{100} =$ _____

11. $\frac{6}{10} + \frac{19}{100} =$ _____

12. $\frac{3}{10} + \frac{35}{100} =$ _____

Lesson 6 Use Place Value and Models to Add

Problem Solving

For Exercises 13 and 14, write each answer as a fraction with a denominator of 100 and as a decimal.

13. Marisa walked her dog $\frac{1}{10}$ mile on Saturday and $\frac{55}{100}$ mile on Sunday. How far did she walk her dog in all?

14. Nevaeh read $\frac{2}{10}$ of a book. Her older sister read $\frac{60}{100}$ of the same book. How much of the book did they read all together?

Brain Builders

15. **Processes & Practices 2** **Use Algebra** Find the unknown in the number sentence $\frac{3}{10} + \frac{\blacksquare}{100} = \frac{41}{100}$. Explain how you can check your answer.

16. **Processes & Practices 6** **Explain to a Friend** Refer to Exercise 15. Write a similar problem. Trade problems with a friend or classmate and solve. Explain how you wrote and solved the problems.

17. **Building on the Essential Question** How does place value help when adding $\frac{1}{10}$ and $\frac{1}{100}$? How could this be used to add $\frac{5}{10}$ and $\frac{5}{100}$?

Name _____

MY Homework

Lesson 6

Use Place Value and Models to Add

Homework Helper

Need help? connectED.mcgraw-hill.com

Julian sold $\frac{6}{10}$ of the school play tickets on Monday. He sold $\frac{23}{100}$ of the tickets on Tuesday. What part of the tickets did Julian sell in all? Write the answer as a fraction with a denominator of 100 and as a decimal.

Use a model to show $\frac{6}{10} + \frac{23}{100}$.

 Write $\frac{6}{10}$ as a fraction with a denominator of 100.

The decimal models show that $\frac{6}{10} = \frac{60}{100}$.

 Add like fractions.

$\frac{60}{100} + \frac{23}{100} = \frac{60 + 23}{100}$

$= \frac{83}{100}$

3 Write the sum as a decimal.

Think of $\frac{83}{100}$ as *eighty-three hundredths*. So, $\frac{83}{100} = 0.83$.

So, Julian sold $\frac{83}{100}$ or 0.83 of the tickets.

Check

The models show that $\frac{6}{10} + \frac{23}{100} = \frac{83}{100}$.

 + =

Lesson 6 My Homework 667

Practice

Add. Write each sum as a fraction with a denominator of 100 and as a decimal.

1. $\frac{2}{10} + \frac{33}{100} = $ _____

2. $\frac{6}{10} + \frac{25}{100} = $ _____

3. $\frac{4}{10} + \frac{17}{100} = $ _____

4. $\frac{2}{10} + \frac{22}{100} = $ _____

Brain Builders

Processes & Practices 2 Use Number Sense Write each answer as a fraction with a denominator of 100 and as a decimal.

5. An insect's body is $\frac{1}{10}$-inch long. The insect's head is $\frac{3}{100}$-inch long. Is the combined length of the insect's body and head longer or shorter than $\frac{4}{10}$ inch? Explain.

_____

_____

6. Makenna rode her bike $\frac{6}{10}$ mile in the morning, $\frac{1}{10}$ mile in the afternoon, and $\frac{23}{100}$ mile in the evening. How far did she ride her bike in all?

_____

7. **Test Practice** Which addition expression describes the model at the right?

Ⓐ $\frac{70}{10} + \frac{18}{100}$ Ⓑ $\frac{7}{10} + \frac{18}{100}$

Ⓒ $\frac{7}{100} + \frac{18}{100}$ Ⓓ $\frac{7}{10} + \frac{18}{10}$

668 **Need more practice?** Download Extra Practice at connectED.mcgraw-hill.com

Name _____

Lesson 7
Compare and Order Decimals

ESSENTIAL QUESTION
How are fractions and decimals related?

To compare decimals, you can use a number line or place value.

 Math in My World

Example 1

At recess, Austin and his friends decided to have two races. One was 0.2 mile, and the other was 0.4 mile. Which race was longer?

One Way Use a number line.

Graph 0.4 and 0.2.

0.4 is to the right of 0.2.

So, 0.4 > 0.2.

Another Way Use place value.

Line up the decimal points. Then compare the digits in each place-value position.

In the tenths place, 4 > 2.

So, 0.4 > 0.2.

So, the longer race was the one that was _____.

Notice that you can only compare 0.4 and 0.2 because they refer to the same whole, which is one mile.

Ones	Tenths
0	2
0	4

Online Content at connectED.mcgraw-hill.com

Lesson 7 669

Example 2
Order 0.15, 0.38, and 0.3 from *greatest* to *least*.

1. Line up the decimal points.

2. Place a zero to the right of 0.3 so all the numbers have the same number of decimal places.

0.15
0.38
0.30

3. Compare the digits in each place-value position.
In the tenths place, 1 < 3.
So, 0.15 is the least decimal. In the hundredths place, 8 > 0. So, 0.38 is the greatest decimal.

The order from *greatest* to *least* is _____, _____, and _____.

Check

0.15   <   0.30   <   0.38

Guided Practice

Compare. Use >, <, or =.

1. 0.2 ◯ 0.6   2. 0.4 ◯ 0.04

3. 0.6 ◯ 0.60   4. 0.57 ◯ 0.7

Order from *greatest* to *least*.

5. 0.2, 0.02, 0.4

_____

6. 0.51, 0.34, 0.46

_____

Explain how to compare 0.4 and 0.40.

670 **Chapter 10** Fractions and Decimals

Name ....

Independent Practice

Compare. Use >, <, or =. Draw a number line to check your work.

7. 0.5 ◯ 0.7   8. 0.56 ◯ 0.58   9. 0.8 ◯ 0.80

10. 0.38 ◯ 0.3   11. 0.90 ◯ 0.9   12. 0.2 ◯ 0.02

13. 0.87 ◯ 0.78   14. 0.48 ◯ 0.5   15. 0.07 ◯ 0.7

Order from *least* to *greatest*.

16. 0.5, 0.55, 0.6   17. 0.35, 0.53, 0.3

_____   _____

18. 0.38, 0.4, 0.36   19. 0.16, 0.6, 0.61

_____   _____

Circle whether each statement is true or false. Use the number line to compare.

20. 0.46 > 0.48   true   false

21. 0.49 > 0.43   true   false

22. 0.42 > 0.47   true   false

23. 0.50 < 0.45   true   false

Lesson 7 Compare and Order Decimals **671**

Problem Solving

24. The distance from Scarlett's house to the elementary school is 0.3 mile. The distance from Scarlett's house to the middle school is 0.9 mile. Which distance is shorter?

25. On a vegetable tray, 0.5 of the vegetables are carrots and 0.2 of the vegetables are broccoli. What vegetable is there more of?

Brain Builders

26. Processes &Practices 2 Use Number Sense What number is halfway between 0.36 and 0.48 on a number line? Explain.

27. Processes &Practices 3 Find the Error Miles wrote the 3 numbers below in order from *least* to *greatest*. Find and correct his mistake.

$$0.27, \frac{19}{100}, 0.29$$

28. Building on the Essential Question How can I use models to compare decimals? Include an example to support your answer.

672 Chapter 10 Fractions and Decimals

Name _____

Lesson 7

Compare and Order Decimals

Homework Helper

Need help? connectED.mcgraw-hill.com

Compare 0.83 and 0.85. Use >, <, or =.

One Way Use a number line.

Graph 0.83 and 0.85.

0.83 is to the left of 0.85.

So, 0.83 < 0.85.

Another Way Use place value.

Line up the decimal points. Then compare the digits in each place-value position.

The digits in the ones places are the same.

The digits in the tenths places are the same.

In the hundredths place, 3 < 5.

So, 0.83 < 0.85.

Ones	Tenths	Hundredths
0	8	3
0	8	5

Practice

Compare. Use >, <, or =.

1. 0.9 ◯ 0.4
2. 0.32 ◯ 0.37
3. 0.9 ◯ 0.90

Order from *least* to *greatest*.

4. 0.7, 0.27, 0.43

5. 0.4, 0.22, 0.72

Lesson 7 My Homework 673

Problem Solving

6. Eli lives 0.7 mile from his school. He lives 0.25 mile from his friend's house. Does Eli live closer to his school or his friend's house?

   _____

7. Alana spent 0.35 of a dollar on a ring at the arcade and 0.72 of a dollar on a snack. Did Alana spend more on the ring or the snack?

   _____

Brain Builders

8. **Processes & Practices 7** **Identify Structure** Tristan has a box of toys. Four out of 100 toys are red, 0.52 of the toys are green, and 0.1 of the toys are blue. Write the color of toys in order from *greatest* to *least*. Explain your answer.

   _____

   _____

9. **Processes & Practices 4** **Model Math** Write a real-world problem that compares 0.4 and 0.29. Then solve the problem and explain your solution.

   _____

   _____

   _____

10. **Test Practice** Which numbers are ordered from *least* to *greatest*?

    Ⓐ 0.25, $\frac{9}{10}$, $\frac{35}{100}$

    Ⓑ $\frac{9}{10}$, 0.25, $\frac{35}{100}$

    Ⓒ 0.25, $\frac{35}{100}$, $\frac{9}{10}$

    Ⓓ $\frac{9}{10}$, $\frac{35}{100}$, 0.25

Name _____

Lesson 8
Problem-Solving Investigation
STRATEGY: Extra or Missing Information

ESSENTIAL QUESTION
How are fractions and decimals related?

Learn the Strategy

On the first day, Gabriella's family traveled $\frac{3}{10}$ of a road trip. On the second day, they traveled $\frac{27}{100}$ of the trip. They traveled 4 days. What part of their trip did they travel in the first two days?

1 Understand

What facts do you know?

Gabriella's family traveled $\frac{3}{10}$ of a road trip on the first day and $\frac{27}{100}$ of the trip on the second day. They traveled 4 days.

What do you need to find?

the part of their trip that they traveled in the first two days

2 Plan

The fact that they traveled 4 days is extra information. Find $\frac{3}{10} + \frac{27}{100}$.

3 Solve

$\frac{3}{10} + \frac{27}{100} = \frac{30}{100} + \frac{27}{100} = \frac{30+27}{100} = \frac{57}{100}$ Write $\frac{3}{10}$ as $\frac{30}{100}$. Then add.

So, Gabriella's family traveled $\frac{\boxed{}}{100}$ of their trip in the first two days.

4 Check

Does your answer make sense? Explain.

_____

Online Content at connectED.mcgraw-hill.com

Lesson 8 675

Practice the Strategy

Charlotte walked $\frac{6}{10}$ mile to school. After school, she walked $\frac{24}{100}$ mile to her friend's house. How much time does it take Charlotte to walk to school and to her friend's house?

1 Understand

What facts do you know?

_____

What do you need to find?

_____

2 Plan

_____

_____

3 Solve

4 Check

Does your answer make sense? Explain.

_____

_____

676 Chapter 10 Fractions and Decimals

Name

Apply the Strategy

Determine if there is extra or missing information to solve each problem. Then solve if possible.

1. There are 100 movies at the store. $\frac{30}{100}$ are action movies, $\frac{50}{100}$ are comedies, and $\frac{20}{100}$ are adventure movies. What part of the movies are action or comedies?

   _____
   _____
   _____

2. **Processes & Practices 1** **Keep Trying** In a basketball game, Jane scored $\frac{3}{10}$ of the baskets in the game and Molly scored $\frac{4}{10}$ of the baskets in the game. There are 10 players on the team. What fraction of the baskets did Jane and Molly score in the game?

   _____
   _____
   _____

Brain Builders

3. Alexia and her family went on vacation. They walked $\frac{1}{10}$ mile to the beach and $\frac{2}{10}$ mile to the souvenir shop. Then they walked back to the beach. How far did they walk in all?

   _____
   _____
   _____

Lesson 8 Problem-Solving Investigation

Review the Strategies

Use any strategy to solve each problem.
- Determine extra or missing information.
- Use logical reasoning.
- Look for a pattern.
- Make a model.

4. Trina is making friendship bracelets. One tenth of the bracelets are blue. Some of the bracelets are red and some are purple. How many bracelets are blue and purple?

_____

_____

5. **Processes & Practices 8** **Look for a Pattern** Find the next number in the pattern below. Explain how you found the number.

$\frac{15}{100}, \frac{3}{10}, \frac{45}{100}, \frac{6}{10}, \frac{75}{100}, \ldots$

_____

_____

_____

6. The fourth grade classes voted on their favorite flavor of ice cream. Three tenths of the students voted for strawberry, $\frac{21}{100}$ of the students voted for vanilla, and $\frac{4}{10}$ of the students voted for chocolate. How many students voted for vanilla or chocolate?

_____

_____

7. Harper and his mom are making trail mix for a party. Two tenths of the trail mix is pretzels and $\frac{32}{100}$ of the trail mix is cereal. The party starts at 1:00 p.m. How much of the trail mix is pretzels or cereal?

_____

_____

678 Chapter 10 Fractions and Decimals

Name _____

Lesson 8

Problem Solving: Extra or Missing Information

Homework Helper   Need help? connectED.mcgraw-hill.com

Stella walked $\frac{4}{10}$ mile to school. During recess, she played soccer. After school, Stella walked $\frac{13}{100}$ mile to the library. How far did Stella walk before and after school?

1 Understand

What facts do you know?

Stella walked $\frac{4}{10}$ mile to school and $\frac{13}{100}$ mile to the library. During recess, she played soccer.

What do you need to find?

the distance Stella walked before and after school

2 Plan

The fact that Stella played soccer at recess is extra information.

Find $\frac{4}{10} + \frac{13}{100}$.

3 Solve

$\frac{4}{10} + \frac{13}{100} = \frac{40}{100} + \frac{13}{100} = \frac{40 + 13}{100} = \frac{53}{100}$ Write $\frac{4}{10}$ as $\frac{40}{100}$. Then add.

So, Stella walked $\frac{53}{100}$ mile.

4 Check

Does your answer make sense? Explain.

Since $\frac{53}{100} - \frac{40}{100} = \frac{13}{100}$, the answer makes sense.

Problem Solving

Determine if there is extra or missing information to solve each problem. Then solve if possible.

1. Janice bought her mother a bunch of 10 flowers. Two of the flowers are daisies. One half of the remaining flowers are tulips. Write the fraction of the flowers that are daisies.

   _____

   _____

2. **Processes & Practices 1 Make a Plan** There are 100 books in the library. There are non-fiction and fiction books. Write the fraction of the books that are fiction.

   _____

   _____

Brain Builders

3. Sean has a collection of coins. One tenth of the coins are from Europe, $\frac{32}{100}$ are from Asia and 0.11 are from South America. The rest are from Africa. Write a decimal to show the total part of the coins that are *not* from Africa.

   _____

   _____

4. Kenley has country, classical, jazz, and pop songs on her digital music player. Of the songs, seventeen hundredths are country, $\frac{2}{10}$ are jazz, and 0.43 are pop songs. She has the same number of classical and jazz songs. What part of the songs are either country or classical? Write as a decimal.

   _____

   _____

Review

Chapter 10

Fractions and Decimals

Vocabulary Check

1. Write three examples of numbers that are **decimals**.

   _____

   Write three examples of numbers that are not **decimals**.

   _____

2. Write three examples of decimals that have a 4 in the **tenths** place.

   _____

3. Write three examples of decimals that have a 5 in the **hundredths** place.

   _____

Concept Check

Write the decimal represented by each model.

4.

   _____

5.

   _____

6.

   _____

My Chapter Review 681

Shade the model. Then write the decimal.

7. four out of ten

 _____

8. thirty-nine out of 100

 _____

9. Write 0.8 as a fraction with a denominator of 10. Shade the grid.

   0.8 = _____

10. Write 0.41 as a fraction with a denominator of 100. Shade the grid.

    0.41 = _____

Write each fraction as an equivalent fraction with a denominator of 100. Then write the fraction as a decimal.

11. $\frac{9}{10} =$ _____

    Decimal: _____

12. $\frac{3}{10} =$ _____

    Decimal: _____

Add. Write each sum as a fraction with a denominator of 100 and as a decimal.

13. $\frac{2}{10} + \frac{36}{100} =$ _____

14. $\frac{7}{10} + \frac{13}{100} =$ _____

Order each set of decimals from *greatest* to *least*.

15. 0.3, 0.23, 0.61 _____

16. 0.72, 0.5, 0.69 _____

Name _____

Problem Solving

17. A barn has 10 animals. Three of the animals are horses. Write a decimal to show what part of the barn animals are horses.

_____

18. Allison bought a snack using the coins shown. Write a decimal to show what part of a dollar she spent.

_____

Brain Builders

19. John lives 0.67 mile from the recreation center. Bella lives 0.8 mile from the recreation center. Maria lives 0.84 mile from the recreation center. Who lives the greatest distance from the recreation center? Who lives the closest? Explain.

_____

20. Joseph ran four tenths mile on Monday, $\frac{35}{100}$ mile on Tuesday, and 0.25 mile on Wednesday. How far did Joseph run in all?

_____

21. Test Practice Chase poured three-tenths liter of lemonade and twenty-nine hundredths liter of raspberry juice into a pitcher. Which decimal is the total amount of liquid Chase poured into the pitcher?

Ⓐ 0.32 liter   Ⓒ 0.69 liter

Ⓑ 0.59 liter   Ⓓ 0.95 liter

My Chapter Review 683

Reflect

Chapter 10
Answering the
ESSENTIAL QUESTION

Use what you learned about fractions and decimals to complete the graphic organizer.

ESSENTIAL QUESTION
How are fractions and decimals related?

Fractions	Both	Decimals

Reflect on the ESSENTIAL QUESTION Write your answer below.

_____

_____

_____

684 Chapter 10 Fractions and Decimals

Name _____   Date _____

Score _____

Performance Task

Wendell's Weather Report

Wendell is in charge of keeping the weather records. He has to record all the rainfall for each local town every day. Some of the reports that come in are written using decimals, and some use fractions. Wendell needs to use both methods in the final report.

Show all your work to receive full credit.

Part A

Complete the table to show the rain totals for Thursday in both fraction and decimal notation.

Location	Alpharetta	Roswell	Kennesaw	Duluth	Marietta
Rain (in.) fraction		$\frac{7}{100}$	$\frac{8}{100}$		$\frac{9}{10}$
Rain (in.) decimal	0.7			0.08	
Location	**Lawrenceville**	**Atlanta**	**Snellville**	**Conyers**	**Covington**
Rain (in.) fraction	$\frac{76}{100}$		$\frac{5}{100}$	$\frac{62}{100}$	
Rain (in.) decimal		0.81			0.8

Online Content at connectED.mcgraw-hill.com   Performance Task 684PT1

Part B

Rank the rain totals from *least* to *greatest* in the table. Indicate locations that are tied and have the same amount of rainfall.

Location	Alpharetta	Roswell	Kennesaw	Duluth	Marietta
Rank					
Location	Lawrenceville	Atlanta	Snellville	Conyers	Covington
Rank					

Part C

Wendell also recorded the rain that fell on Friday at each location. What are the totals for the two days? Complete the table.

Location	Alpharetta	Roswell	Kennesaw	Duluth	Marietta
Rain (in.) Thursday	0.7	0.07	0.08	0.08	0.9
Rain (in.) Friday	0.14	0.28	0.17	0.34	0.05
Total (in.) Thurs/Fri					
Location	Lawrenceville	Atlanta	Snellville	Conyers	Covington
Rain (in.) Thursday	0.76	0.81	0.05	0.62	0.8
Rain (in.) Friday	0.17	0.06	0.43	0.2	0.16
Total (in.) Thurs/Fri					

Chapter 11 Customary Measurement

ESSENTIAL QUESTION
Why do we convert measurements?

Let's Discover Nature!

Watch a video!

Name
....

MY Chapter Project

Healthful Snacks

1. With your partner, find a recipe online for a healthful snack.

Our recipe is: _____

_____

_____

2. List the ingredients needed for your recipe in the table below. Estimate how much of each ingredient is needed to make enough of the recipe for the whole class. Then convert the estimated amount of each ingredient into the largest unit of capacity or weight possible. For example, 4 cups of peaches = 2 pints of peaches. Use another sheet of paper, if needed.

Ingredient	Estimated Amount	Amount in Largest Unit

3. Make a poster showing your healthful snack recipe with the amounts of each ingredient that are needed for the whole class.

_____

_____

686 Chapter 11 Customary Measurement

Name _____

Am I Ready?

Circle the longer figure.

1. Figure A ──────────────────
   Figure B ──────────

2. Figure A [ ]
   Figure B [ ]

Multiply.

3. 12
   × 9

4. 16
   × 5

5. 225
   × 5

Add. Write in simplest form.

6. $\frac{2}{10} + \frac{5}{10} =$ _____

7. $\frac{1}{4} + \frac{1}{4} =$ _____

8. $\frac{1}{6} + \frac{3}{6} =$ _____

Write the time shown on each clock.

9.
_____

10.
_____

11.
_____

Shade the boxes to show the problems you answered correctly.

How Did I Do? | 1 | 2 | 3 | 4 | 5 | 6 | 7 | 8 | 9 | 10 | 11 |

Online Content at connectED.mcgraw-hill.com

Name ....

MY Math Words

Review Vocabulary

estimate   length   time   weight

Making Connections
Use review vocabulary to write a sentence describing how you would find each measurement. Underline the word. Then provide an estimate for each question.

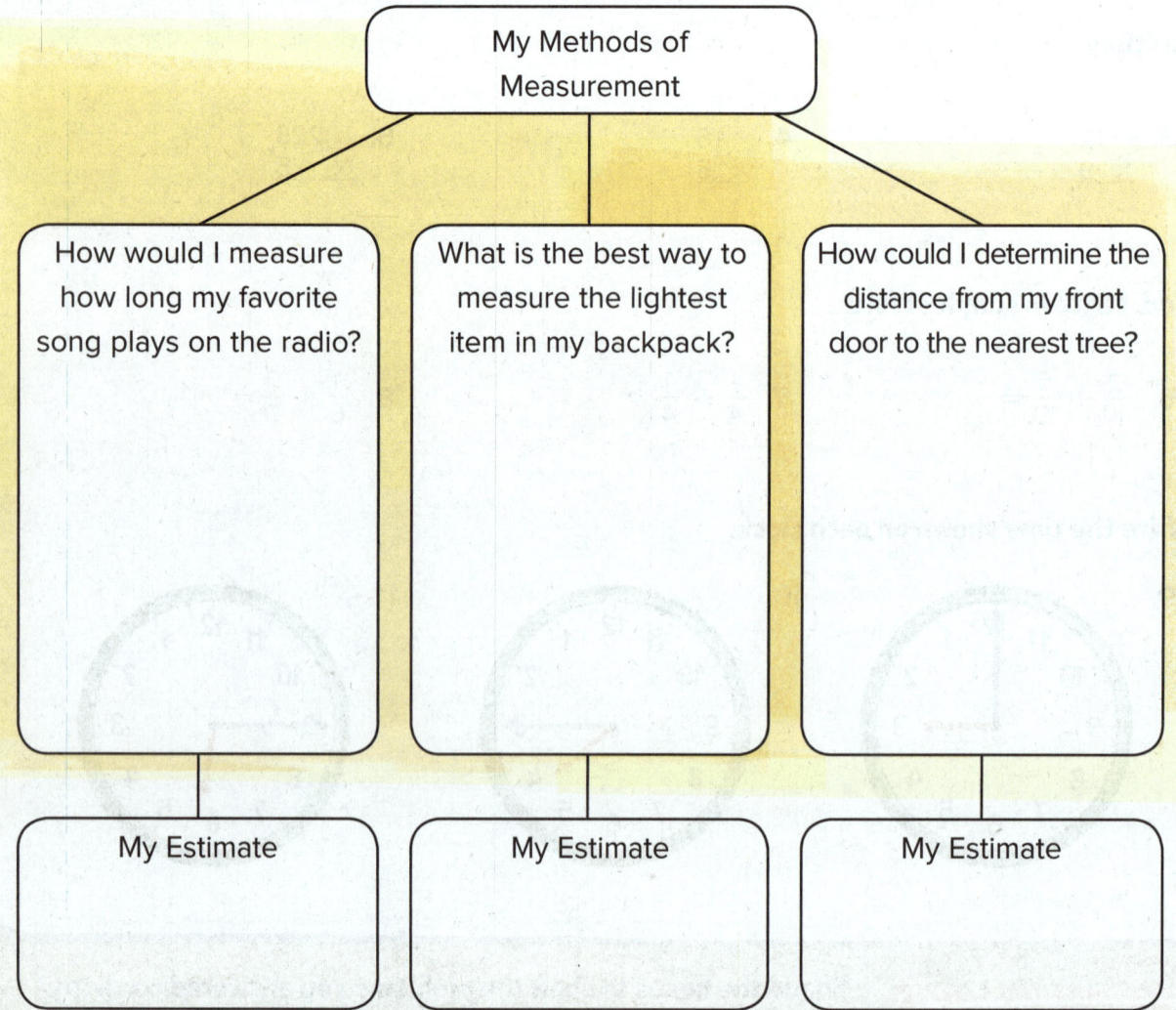

My Methods of Measurement

- How would I measure how long my favorite song plays on the radio?
  - My Estimate
- What is the best way to measure the lightest item in my backpack?
  - My Estimate
- How could I determine the distance from my front door to the nearest tree?
  - My Estimate

MY Vocabulary Cards

Lesson 11–3

capacity

Lesson 11–2

convert

? quarts = 5 gallons
4 quarts = 1 gallon
4 × 5 = 20 quarts
20 quarts = 5 gallons

? feet = 156 inches
1 foot = 12 inches
13 × 12 = 156
13 feet = 156 inches

Lesson 11–3

cup (c)

1 cup

Lesson 11–1

customary system

1 pound 5 ounces

Lesson 11–3

fluid ounce (fl oz)

8 fluid ounces

Lesson 11–1

foot (ft)

1 foot

Lesson 11–3

gallon (gal)

1 gallon

Lesson 11–8

line plot

Ideas for Use

- Group 2 or 3 common words. Add a word that is unrelated to the group. Then work with a friend to name the unrelated word.
- Design a crossword puzzle. Use the definitions for the words as the clues.

To change to different units.
Which operation would you use to find the number of feet in 2 miles? Solve.

The amount a container can hold.
Write a sentence describing the capacity of a bathtub.

The units of measurement most often used in the United States, such as the inch, yard, and mile.
What customary unit of measurement would you use to describe the distance from California to New York?

A customary unit of capacity.
Write a sentence using the multiple-meaning word *cup* as a verb.

A customary unit for measuring length.
This unit of length is based on the length of an adult's foot. How could this information be useful?

A customary unit of capacity.
Give examples of two things at the grocery store that are measured in fluid ounces.

A diagram showing the frequency of data on a number line.
Describe what the line plot on the front of the card might represent.

A customary unit of capacity.
Write a riddle to help you remember how many quarts, pints, and cups are in a gallon.

MY Vocabulary Cards

Lesson 11–2

mile (mi)

Lesson 11–5

ounce (oz)

1 ounce

Lesson 11–3

pint (pt)

1 pint

Lesson 11–5

pound (lb)

1 pound

Lesson 11–3

quart (qt)

1 quart

Lesson 11–7

second (s)

60 seconds = 1 minute

Lesson 11–5

ton (T)

1 ton

Lesson 11–5

weight

Ideas for Use

- Write a tally mark on each card every time you read the word in this chapter or use it in your writing. Challenge yourself to make at least 10 tally marks for each card.

- During this school year, create a separate stack of cards for key math verbs, such as *convert*. These verbs will help you in your problem solving.

A customary unit for measuring weight.
An 8-week old kitten weighs about 32 ounces. Explain if it is reasonable to weigh an adult cat in ounces.

A customary unit for measuring length.
It takes Tamika about 12 minutes to walk 1 mile. About how many minutes would it take Tamika to walk 3 miles?

A customary unit for measuring weight.
Estimate the weight of one of your textbooks to the nearest pound. Explain your reasoning.

A customary unit for measuring capacity.
1 pint = 2 cups
How many pints are in a quart? In a gallon?

A unit of time.
60 seconds = 1 minute
If there are 60 seconds in 1 minute, how many seconds are in 10 minutes?

A customary unit for measuring capacity.
Is 6 quarts greater than, less than, or equal to 2 gallons? Explain.

A measurement that tells how heavy an object is.
Homophones are words that sounds alike, but are spelled differently. New and knew are homophones. Write a homophone for the word weight and its definition.

A customary unit for measuring weight.
What unit of weight would you use to measure a school desk? Explain.

MY Vocabulary Cards

Lesson 11–1

yard (yd)

1 yard

Ideas for Use

- Pick a few lessons from the chapter. Write the name of each lesson on the front of the blank cards. Write a few study tips for each lesson on the back of each card.

- Have students use the blank cards to write summaries of the conversions they learned in the chapter.

A customary unit for measuring length.

Give examples of three things you might measure in yards.

_____
_____
_____

MY Foldable

FOLDABLES Follow the steps on the back to make your Foldable.

My Customary Conversions

Weight
Length
Capacity

Capacity

	fluid ounce	cup (c)
fluid ounce (fl oz)	1 fluid ounce	___ fl oz = 1 c
cup (c)	1 c = ___ fl oz	1 cup
pint (pt)	1 pt = 16 fl oz	1 pt = ___ c
quart (qt)	1 qt = 32 fl oz	___ qt = 4 c
gallon (gal)	1 gal = 128 fl oz	1 gal = 16 c

pound (lb)	ton (T)
16 oz = ___ lb	32,000 oz = 1 T
1 pound	___ lb = 1 T
1 T = ___ lb	1 Ton

FOLDABLES® Study Organizer

Weight

	ounce (oz)
ounce (oz)	1 ounce
pound (lb)	1 lb = ___ oz
ton (T)	1 T = 32,000 oz

	pint (pt)	quart (qt)	gallon (gal)
	16 fl oz = 1 pt	32 fl oz = 1 qt	128 fl oz = 1 gal
	___ c = 1 pt	___ c = 1 qt	___ c = 1 gal
	1 pint	___ pt = 1 qt	___ pt = 1 gal
	___ qt = 2 pt	1 quart	___ qt = 1 gal
	1 gal = 8 pt	___ gal = 4 qt	1 gallon

Length

	inch (in.)	foot (ft)	yard (yd)	mile (mi)
inch (in.)	1 inch	12 in. = ___ ft	___ in. = 1 yd	63,360 in. = 1 mi
foot (ft)	1 ft = ___ in.	1 foot	___ ft = 1 yd	___ ft = 1 mi
yard (yd)	___ yd = 36 in.	___ yd = 3 ft	1 yard	1,760 yd = 1 mi
mile (mi)	1 mi = 63,360 in.	___ mi = 5,280 ft	1 mi = 1,760 yd	1 mile

Name _____

Lesson 1
Customary Units of Length

ESSENTIAL QUESTION
Why do we convert measurements?

Length is the measurement of a line between two points. Inch, foot, and yard are units that are part of the **customary system** of measure for length.

Math in My World

Example 1

The actual size of a neon damsel marine fish is shown. Measure how long the fish is to the nearest inch, $\frac{1}{2}$ inch, and $\frac{1}{4}$ inch.

 Measure to the nearest inch.

Compare the length to what you know about inches. Record the length in the table.

 Measure to the nearest $\frac{1}{2}$ inch.

Using a ruler, measure the length of the fish to the nearest $\frac{1}{2}$ inch. Record the length in the table.

 Measure to the nearest $\frac{1}{4}$ inch.

Measure the length of the fish to the nearest $\frac{1}{4}$ inch. Record the length in the table.

Length of Neon Damsel Marine Fish	
Nearest Inch	
Nearest $\frac{1}{2}$ inch	
Nearest $\frac{1}{4}$ inch	

An inch (in.) is about the length of one paper clip.

A **foot (ft)** is about the length of a textbook.

A **yard (yd)** is about the height of a chair.

Online Content at connectED.mcgraw-hill.com

Lesson 1 697

Example 2

Choose the best estimate for the length of the caterpillar.

Ⓐ 2 inches

Ⓑ 20 inches

Ⓒ 20 feet

Ⓓ 2 yards

A caterpillar is small. So, inches are better estimates than feet or yards. Since 20 inches is more than a foot and is too big, the best estimate is _____.

Guided Practice

Estimate. Then measure each to the nearest inch, $\frac{1}{2}$ inch, and $\frac{1}{4}$ inch.

1.

2.

_____   _____

3. Choose the best estimate for the length.

   Ⓐ 12 inches

   Ⓑ 4 feet

   Ⓒ 12 feet

   Ⓓ 4 yards

Talk MATH

Why do you think there is more than one unit of length for measure?

Name ....

Independent Practice

Estimate. Then measure each to the nearest inch, $\frac{1}{2}$ inch, and $\frac{1}{4}$ inch.

4.

   _____

5.

   _____

6.

   _____

7.

   _____

Choose the best estimate for each length.

8. length of a whistle

   Ⓐ 2 yards
   Ⓑ 2 feet
   Ⓒ 12 inches
   Ⓓ 2 inches

9. length of a chalkboard

   Ⓕ 1 foot
   Ⓖ 2 feet
   Ⓗ 1 yard
   Ⓘ 2 yards

Lesson 1 Customary Units of Length 699

Problem Solving

10. Patrice found a book that was $2\frac{1}{4}$ inches thick. She stacked it with another book that was the same thickness. How tall was the stack?

11. Helki found a stick that was $5\frac{3}{4}$ feet long. He needed one that was about 5 feet long. Is it reasonable to say that this stick will work? Explain.

Brain Builders

12. Processes & Practices 4 Model Math Use the > and < signs to put the following units in order from least to greatest: foot, yard, inch.

Explain your answer.

13. Processes & Practices 5 Use Math Tools Find two objects in your classroom that are longer than two inches and shorter than 4 inches. How did you use estimation in selecting objects?

14. Building on the Essential Question Name two customary units of length. Give an example of an object and tell which unit is more accurate for measuring that object. Explain.

700 Chapter 11 Customary Measurement

Name _____

MY Homework

Lesson 1
Customary Units of Length

Homework Helper

Need help? connectED.mcgraw-hill.com

Estimate the length. Then measure to the nearest inch, $\frac{1}{2}$ inch, and $\frac{1}{4}$ inch.

1 Estimate.
Compare the length of the crayon to what you know about inches. You know that a paper clip is about 1 inch long. You can estimate that the crayon is about 3 paper clips, or 3 inches, long.

$\frac{1}{4}$ inches

2 Measure to the Nearest:

Inch: The length is closer to 3 inches than 4 inches.

$\frac{1}{2}$ **inch:** The length is closer to $3\frac{1}{2}$ inches than to 3 inches.

$\frac{1}{4}$ **inch:** The length is closer to $3\frac{1}{4}$ inches than to $3\frac{1}{2}$ inches.

So, the crayon is about 3 inches, $3\frac{1}{2}$ inches, or $3\frac{1}{4}$ inches long.

Practice

Estimate. Then measure each to the nearest inch, $\frac{1}{2}$ inch, and $\frac{1}{4}$ inch.

1.

2.

Lesson 1 My Homework 701

Which customary unit—inch, foot, or yard—is best for measuring the length of each of the following?

3. baseball _____
4. refrigerator _____
5. football field _____
6. jump rope _____
7. parking lot _____
8. shoe _____

Problem Solving

9. Brice needs 2 feet of ribbon to tie a bow around a birthday present. He has 1 yard of ribbon. Is that enough?

10. Clark says that his parents' car is 96 units long. Jay says that his parents' car is 8 units long. Which customary unit of length was each person using?

Brain Builders

11. **Processes & Practices** **Check for Reasonableness** Glenna says that the height of her house is 1 yard. Is her claim reasonable? Explain.

Vocabulary Check

Write a vocabulary term in the correct space to complete each sentence.

    customary   foot   yard

12. An inch is one kind of _____ unit for measuring length.
13. A textbook is about 1 _____ long.
14. A door is about 1 _____ wide.
15. **Test Practice** Which is the best estimate for the length of four pencils lined up end to end?
    Ⓐ 2 inches   Ⓑ 2 feet   Ⓒ 2 yards   Ⓓ 20 yards

Name _____

Lesson 2
Convert Customary Units of Length

ESSENTIAL QUESTION
Why do we convert measurements?

You can multiply to **convert**, or change between, units.

Customary Units of Length
1 foot (ft) = 12 inches (in.)
1 yard (yd) = 3 feet (ft)
1 **mile (mi)** = 5,280 feet (ft)

Math in My World

Example 1

Marla's dog, Cory, loves to jump into water. Cory's longest jump is 7 yards. How many feet are in 7 yards?

You know the number of yards and want to find the number of feet. Yards are a larger unit than feet. One yard is 3 times as long as one foot. So, use multiplication.

7 yd = ? ft

Multiply by 3 because 1 yard = 3 feet.   7 × 3 = _____

So, there are _____ feet in 3 yards.

Check

You can check by drawing a picture. You know that 3 feet = 1 yard.

Online Content at **connectED.mcgraw-hill.com**

Lesson 2 703

Example 2

Convert $3\frac{1}{2}$ feet to inches.

$3\frac{1}{2}$ ft = ? in.

 Convert the whole number.

One foot is 12 times as long as one inch. So, use multiplication.

$3 \times 12 =$ _____

 Convert the fraction.

Since 1 foot = 12 inches, half of a foot equals half of 12, or 6 inches.

 Add the parts.

36 inches + 6 inches = 42 inches

So, there are _____ inches in $3\frac{1}{2}$ feet.

Guided Practice

Complete each conversion table to show measurement equivalents.

1.

inches (in.)	feet (ft)	(in., ft)
	1	(12, 1)
	2	(24, 2)
	3	
	4	

List the number pairs in the last column of the conversion table.

2.

yards (yd)	inches (in.)	(yd, in.)
	36	
	108	
	216	
	252	

Talk MATH

When converting from larger units to smaller units, do you need to multiply or divide?

Name _____

Independent Practice

Complete each conversion table.

3.
yards (yd)	feet (ft)	(yd, ft)
2		
3		
4		
5		

4.
yards (yd)	inches (in.)	(yd, in.)
2		
4		
5		
8		

5.
feet (ft)	inches (in.)	(ft, in.)
3		
6		
9		
12		

6.
feet (ft)	miles (mi)	(ft, mi)
	1	
	2	
	3	
	4	

Algebra Find each unknown number.

7. 11 ft = ■ in.

   ■ = _____

8. 18 yd = ■ ft

   ■ = _____

9. $2\frac{1}{3}$ yd = ■ ft

   ■ = _____

10. How many times longer is one yard than one foot? _____

11. How many times longer is one mile than one foot? _____

12. How many times shorter is one inch than one foot? _____

13. How many times shorter is one foot than one yard? _____

Lesson 2 Convert Customary Units of Length 705

Problem Solving

14. Ramiro sits $5\frac{1}{2}$ feet from the bookshelf. Michelle sits 64 inches from the bookshelf. Who sits closer to the bookshelf?

15. The Costa family hiked a trail that was 2 miles in one direction. How many feet was the hike round-trip?

16. Processes & Practices 3 Justify Conclusions Sumi lives 2 miles from school. Valerie lives 10,542 feet from school. Who lives closer to school? Explain your answer.

Brain Builders

17. Processes & Practices 1 Make Sense of Problems Darin is 4 feet 10 inches tall. His brother is 68 inches tall. How many inches taller is Darin's brother than Darin? Explain.

18. Processes & Practices 4 Model Math Write a real-world problem involving the conversion of customary lengths. Explain the solution.

19. Building on the Essential Question How are inches, yards, and feet related?

706 Chapter 11 Customary Measurement

Name ....

MY Homework

Lesson 2

Convert Customary Units of Length

Homework Helper

Need help? connectED.mcgraw-hill.com

Andrew lives 2 miles from the school. How many feet from the school does Andrew live?

To convert miles to feet, multiply the number of miles measured by the number of feet in 1 mile. You know that 1 mile = 5,280 feet.

→ 2 miles = 10,560 feet

So, Andrew lives 10,560 feet from the school.

Customary Units of Length
1 foot (ft) = 12 inches (in.)
1 yard (yd) = 3 feet (ft)
1 mile (mi) = 5,280 feet (ft)

Conversion table		
miles (mi)	feet (ft)	(mi, ft)
2	10,560	(2, 10,560)
4	21,120	(4, 21,120)
6	31,680	(6, 31,680)
8	42,240	(8, 42,240)

Practice

Convert units to complete each equation.

1. 3 ft = _____ in.

2. $6\frac{1}{2}$ ft = _____ in.

3. 4 yd = _____ in.

4. 760 yd = _____ ft

5. 7 ft = _____ in.

6. 3 yd = _____ in.

7. $2\frac{1}{3}$ yd = _____ ft

8. 5 ft = _____ in.

9. 64 ft = _____ in.

10. _____ in. = 6 yd

11. _____ ft = 7 yd

12. $12\frac{1}{2}$ ft = _____ in.

Draw lines to match the equivalent lengths.

13. 24 ft
14. 120 in.
15. 2 yd
16. 5 yd
17. 20 yd

- 60 ft
- 72 in.
- 180 in.
- 8 yd
- 10 ft

Problem Solving

18. **Processes & Practices 5** **Use Math Tools** The Millers' house is 25 yards from Mrs. Shapiro's house. How many feet apart are the two houses?

_____

19. Kate walks half of a mile to the library. How many yards does she walk?

_____

Brain Builders

20. Pe Ling needs 2 yards of fabric for a craft project. She has 3 feet already. How many inches of fabric does she still need? Explain.

_____
_____

Vocabulary Check

Write a vocabulary term in the correct space to complete each sentence.

convert   mile

21. One _____ equals 5,280 feet.

22. To _____ from feet to inches, multiply by 12.

23. **Test Practice** Matthew is 2 yards tall. How many inches tall is he?

Ⓐ 74 inches   Ⓑ 72 inches   Ⓒ 64 inches   Ⓓ 60 inches

Name

Lesson 3
Customary Units of Capacity

ESSENTIAL QUESTION
Why do we convert measurements?

The amount of liquid a container can hold is its **capacity**. Different containers measure different capacities.

1 fl oz   1 cup   1 pint   1 quart   1 gallon

 Math in My World

Example 1

Jorge is filling an aquarium. Which container should Jorge use to fill his aquarium most quickly?

1 cup   1 quart   1 gallon

To fill the aquarium most quickly, Jorge should use the container that will hold the most liquid.

The _____ is the largest unit. It will fill the aquarium most quickly.

Online Content at connectED.mcgraw-hill.com

Lesson 3   709

Example 2

Nita is pouring salsa into a small bowl. Is the most reasonable estimate for the capacity of the bowl: 8 fluid ounces, 8 cups, 8 quarts, or 8 gallons?

The salsa is a small amount. So, 8 gallons, 8 quarts, and 8 cups are too large.

The most reasonable estimate for the capacity of the bowl is _____.

Guided Practice

Choose the most reasonable estimate for each capacity.

1.

   Ⓐ 1 fluid ounce
   Ⓑ 1 pint
   Ⓒ 1 quart
   Ⓓ 100 quarts

2.

   Ⓕ 4 fluid ounces
   Ⓖ 10 cups
   Ⓗ 400 cups
   Ⓘ 10 gallons

3.

   Ⓐ 1 fluid ounce
   Ⓑ 1 cup
   Ⓒ 1 pint
   Ⓓ 1 gallon

Talk MATH

Is it possible for two containers with different shapes to have a capacity of 1 pint? Explain.

710 Chapter 11 Customary Measurement

Name ....

Independent Practice

Choose the most reasonable estimate for each capacity.

4.

   Ⓐ 12 fluid ounces
   Ⓑ 1 gallon
   Ⓒ 12 gallons
   Ⓓ 12,000 gallons

5.

   Ⓕ 2 fluid ounces
   Ⓖ 2 cups
   Ⓗ 2 pints
   Ⓘ 2 gallons

6.

   Ⓐ 1 fluid ounce
   Ⓑ 1 cup
   Ⓒ 1 quart
   Ⓓ 1 gallon

7.

   Ⓕ 8 fluid ounces
   Ⓖ 8 cups
   Ⓗ 8 pints
   Ⓘ 8 gallons

8.

   Ⓐ 1 quart
   Ⓑ 10 quarts
   Ⓒ 100 quarts
   Ⓓ 1,000 quarts

9.

   Ⓕ 16 gallons
   Ⓖ 16 quarts
   Ⓗ 16 fluid ounces
   Ⓘ 16 cups

Estimate the capacity of each object.

10. water bottle

    _____

11. juice box

    _____

12. sink

    _____

13. Is it faster to water two large flower pots using a 1-cup pitcher or a 1-quart pitcher? _____

Lesson 3 Customary Units of Capacity **711**

Problem Solving

Some household activities and the amount of water they consume are listed in the table. Use the table for Exercises 14 and 15.

Water Consumption	
Activity	Water Used (gallons)
Take shower	15–30
Brush teeth (water running)	1–2
Wash dishes (by hand)	20
Wash dishes (in dishwasher)	9–12
Flush toilet	5–7

14. If Callie takes one shower each day, is it reasonable to say that she could use 210 gallons of water in one week? Explain.

_____

_____

15. Processes & Practices 3 **Draw a Conclusion** Callie brushes her teeth three times each day. She leaves the water running. Is it reasonable to say that she uses 2 cups of water in one day? Explain.

_____

_____

Brain Builders

16. Processes & Practices 2 **Reason** Name two things in your classroom that can hold more than 1 cup. Estimate the capacity of each.

_____

17. Processes & Practices 3 **Find the Error** Alexander estimates the capacity of a small soup can to be about 1 fluid ounce. Find and correct his mistake.

_____

_____

18. Building on the Essential Question What is the difference between length and capacity?

_____

_____

712 Chapter 11 Customary Measurement

Name

MY Homework

Lesson 3

Customary Units of Capacity

Homework Helper

Need help? connectED.mcgraw-hill.com

Ethan is pouring himself a drink of juice. Is it reasonable to say that he will pour about a gallon of juice? Explain.

You know that a large carton of milk is a gallon. That would be way too much for one drink. It is not reasonable to say that Ethan will pour about a gallon of juice.

One cup equals 8 fluid ounces. It would be more reasonable to say that Ethan is pouring one cup, or 8 fluid ounces, of juice.

Practice

Choose the best estimate for each capacity.

1.
   - Ⓐ 1 fluid ounce
   - Ⓑ 4 fluid ounces
   - Ⓒ 1 cup
   - Ⓓ 4 cups

2.
   - Ⓕ 6 fluid ounces
   - Ⓖ 16 fluid ounces
   - Ⓗ 1 gallon
   - Ⓘ 6 gallons

3.
   - Ⓐ 50 gallons
   - Ⓑ 50 pints
   - Ⓒ 50 cups
   - Ⓓ 50 ounces

4.
   - Ⓕ 8 quarts
   - Ⓖ 8 pints
   - Ⓗ 8 cups
   - Ⓘ 8 fluid ounces

Draw lines to match each object to a reasonable capacity.

5. cereal bowl • 4 fluid ounces

6. baby food jar • 2 cups

7. soda can • 1,000 gallons

8. swimming pool • 12 fluid ounces

9. large can of paint • 1 gallon

Brain Builders

10. Zach is filling his dog's bowl with water. Is it reasonable to say that Zach will need about 4 gallons of water? Explain.

_____

11. **Processes &Practices** **Justify Conclusions** Wes is pouring himself some ketchup to eat with his potato wedges. What is a reasonable estimate for the number of fluid ounces of ketchup Wes will need? Explain.

_____

Vocabulary Check

12. Explain one relationship between capacity and fluid ounces.

_____
_____

13. Order the following units of capacity from *least* to *greatest*: gallon, pint, cup, quart

_____
_____

14. **Test Practice** Henry gargles with mouthwash twice a day. Which is a reasonable amount of mouthwash for Henry to use each day?

Ⓐ 2 fluid ounces   Ⓒ 12 fluid ounces

Ⓑ 8 fluid ounces   Ⓓ 20 fluid ounces

Name _____

Lesson 4
Convert Customary Units of Capacity

ESSENTIAL QUESTION
Why do we convert measurements?

You can use multiplication to convert units. To change from a larger unit to a smaller unit, multiply.

Customary Units of Capacity	
1 cup (c) = 8 fluid ounces (fl oz)	2 pints (pt) = 1 quart (qt)
2 cups (c) = 1 pint (pt)	4 quarts (qt) = 1 gallon (gal)

Math in My World

Example 1

Marcus has a 2-gallon container of laundry detergent. How many quarts of laundry detergent does he have? How many pints of laundry detergent does he have?

1 **Find the number of quarts that are in 2 gallons.**

Since quarts are smaller than gallons, multiply. Multiply by 4 because there are 4 quarts in each gallon.

2 × 4 = _____

So, there are _____ quarts in 2 gallons.

2 **Find the number of pints that are in 8 quarts.**

Multiply 8 by 2 because there are 2 pints in each quart.

8 × 2 = _____

So, there are _____ pints in 8 quarts.

Marcus has _____ quarts, or _____ pints, of laundry detergent.

Online Content at connectED.mcgraw-hill.com

Lesson 4 715

Example 2

Complete. $4\frac{1}{2}$ quarts = ■ pints

Helpful Hint
When converting measures, there will be more smaller units than larger units.

 Convert the whole number.

There are 2 pints in one quart.

Since pints are smaller than quarts, multiply. $4 \times 2 =$ _____

 Convert the fraction.

Multiply. $\frac{1}{2} \times 2 =$ _____ ← Half of 2 pints is 1 pint.

 Add the parts.

$8 + 1 =$ _____

So, $4\frac{1}{2}$ quarts = _____ pints.

Guided Practice

Complete.

1. 10 qt = _____ pt
2. $3\frac{1}{2}$ c = _____ fl oz
3. 5 qt = _____ pt

Complete each conversion table.

4.

pints (pt)	cups (c)	(pt, c)
1		(1, 2)
2		
3		
4		

Talk MATH

Explain how to convert 6 pints to cups.

5.

gallons (gal)	quarts (qt)	(gal, qt)
2		
4		
6		
8		

716 Chapter 11 Customary Measurement

Independent Practice

Complete each conversion table.

6.

quarts (qt)	pints (pt)	(qt, pt)
1		
2		
3		
4		

7.

pints (pt)	cups (c)	(pt, c)
5		
7		
9		
11		

Algebra Find each unknown number.

8. $8 \text{ c} = \blacksquare \text{ fl oz}$

    $\blacksquare =$ _____

9. $6\frac{1}{2} \text{ gal} = \blacksquare \text{ qt}$

    $\blacksquare =$ _____

10. $\blacksquare \text{ qt} = 5 \text{ gal}$

    $\blacksquare =$ _____

11. $5\frac{1}{2} \text{ c} = \blacksquare \text{ fl oz}$

    $\blacksquare =$ _____

12. $\blacksquare \text{ c} = 15 \text{ pt}$

    $\blacksquare =$ _____

13. $16 \text{ c} = \blacksquare \text{ fl oz}$

    $\blacksquare =$ _____

Compare. Use >, <, or =.

14. 4 qt ◯ 10 pt

15. 10 gal ◯ 1,280 fl oz

16. 1 qt ◯ 2 c

17. 1 gal ◯ 16 c

18. 5 qt ◯ 25 c

19. 12 fl oz ◯ 2 c

20. How many times greater is the capacity of one gallon than one quart? _____

21. How many times greater is the capacity of one cup than one fluid ounce? _____

Lesson 4 Convert Customary Units of Capacity 717

Problem Solving

22. Lucia is making 2 gallons of soup. How many cups of soup is Lucia making?

_____

23. **Processes & Practices 5** **Use Math Tools** Tomas is buying a 2-cup container of liquid dish soap. How many fluid ounces of dish soap is he buying?

_____

24. Danielle is using 2 quarts of water in a recipe. How many cups of water is she using?

_____

Brain Builders

25. Karen is buying 4 gallons of orange juice and 4 pints of apple juice. How many total quarts of juice is she buying?

_____

26. **Processes & Practices 3** **Which One Doesn't Belong?** Circle the measurement that does not belong with the other three. Explain your reasoning and include a measurement that would belong.

| 4 pints | 2 quarts | 8 cups | 1 gallon |

_____

27. **Building on the Essential Question** How are gallons and fluid ounces related?

_____
_____
_____

718 Chapter 11 Customary Measurement

Name

MY Homework

Lesson 4

Convert Customary Units of Capacity

Homework Helper

Need help? connectED.mcgraw-hill.com

Chip needs 1 quart of cream for his ice cream recipe. He has 3 cups of cream. Is that enough for his recipe?

First, convert quarts to pints.

1 quart = 2 pints

Next, convert pints to cups.

2 × 2 = 4 cups

(number of pints) (number of cups in each pint)

Conversion Table Customary Units of Capacity
1 cup (c) = 8 fluid ounces (fl oz)
2 cups (c) = 1 pint (pt)
2 pints (pt) = 1 quart (qt)
4 quarts (qt) = 1 gallon (gal)

Chip has 3 cups of cream, but he needs 4 cups. He does not have enough cream for the recipe.

Practice

Convert units to complete each equation.

1. 9 pt = _____ c

2. 4 qt = _____ pt

3. 3 c = _____ fl oz

4. $9\frac{1}{2}$ gal = _____ qt

5. $5\frac{1}{2}$ pt = _____ c

6. 6 gal = _____ qt

Lesson 4 My Homework 719

Complete each conversion table.

7.

cups (c)	fluid ounces (fl oz)	(c, fl oz)
1		
3		
5		
7		

8.

quarts (qt)	pints (pt)	(qt, pt)
1		
2		
3		
4		

9.

gallons (gal)	quarts (qt)	(gal, qt)
4		
5		
6		
7		

10.

gallons (gal)	pints (pt)	(gal, pt)
4		
5		
6		
7		

Brain Builders

11. **Processes & Practices 2** **Reason** Elle can buy 2 quarts of milk for $3 or 1 gallon of milk for $3. Which is the better deal? Explain.

12. Marco needs 2 gallons of paint for his living room. He has 10 quarts of paint. Is that enough? Explain.

13. Ollie drank $\frac{1}{2}$ of a 16-fluid ounce bottle of juice. How many cups of juice are left in the bottle? Explain.

14. **Test Practice** Craig drank two 5-cup bottles of water in one day. How many fluid ounces is that?

Ⓐ 10 fluid ounces   Ⓒ 40 fluid ounces

Ⓑ 16 fluid ounces   Ⓓ 80 fluid ounces

Check My Progress

Vocabulary Check

Circle the word(s) that completes each sentence.

1. Foot, cup, and gallon are all units in the (**customary system**   capacity).

2. There are 2 (quarts   **cups**) in a pint.

3. In order to (**convert**   pint) yards to feet, multiply the number of yards by 3.

4. Quart and gallon are units of (**capacity**   yards).

5. There are 5,280 feet in 1 (yard   **mile**).

Draw lines to match the measurements that are equal.

6. 1 **foot** •   • 16 **quarts**

7. 32 **fluid ounces** •   • 12 **inches**

8. 4 **gallons** •   • 2 **pints**

Concept Check

Complete each conversion table.

9.

yards (yd)	feet (ft)	(yd, ft)
1		
3		
5		
7		

10.

quarts (qt)	cups (c)	(qt, c)
2		
3		
4		
5		

Problem Solving

11. Mateo is making 2 gallons of fruit punch. How many pints of fruit punch is he making?

_____

12. Gwenith has 3 gallons of milk. How many quarts of milk does she have?

_____

Brain Builders

13. One male boa constrictor is 10 feet long. A female boa constrictor is 4 yards long. Which boa constrictor is longer? Explain.

_____

14. Test Practice Which of the following holds about 2 pints of water?

Ⓐ

Ⓑ

Ⓒ

Ⓓ

Name _____

Lesson 5
Customary Units of Weight

ESSENTIAL QUESTION
Why do we convert measurements?

The **weight** of an object is how heavy it is. The customary units of weight are ounce (oz), pound (lb), and ton (T).

1 ounce (oz)   1 pound (lb)   1 ton (T)

Math in My World

Example 1

Suzie's father bought some sugar for their favorite recipe. Which is a more reasonable unit to use for the weight of a bag of sugar, ounces or pounds?

A small packet of sugar would be weighed in ounces.

A bag of sugar is much larger and would be weighed in pounds.

So, _____ is a reasonable unit to use for the weight of a bag of sugar.

Online Content at connectED.mcgraw-hill.com

Example 2

Which is the most reasonable estimate for the weight of a leaf: 1 ounce, 1 pound, 1 ton, or 10 tons?

Compare the weight of a leaf to the weight of objects that you know.

A leaf weighs less than a pineapple, or 1 pound.

Objects that weigh less than 1 pound are weighed in ounces. The only option that contains ounces is 1 ounce.

So, a leaf weighs about _____.

Guided Practice

Choose the best estimate for the weight of each object.

1. paper airplane

- Ⓐ 4 ounces
- Ⓑ 40 ounces
- Ⓒ 4 pounds
- Ⓓ 4 tons

2. helicopter

- Ⓕ 5 ounces
- Ⓖ 500 ounces
- Ⓗ 5 tons
- Ⓘ 500 tons

3. rabbit

- Ⓐ 4 ounces
- Ⓑ 4 pounds
- Ⓒ 40 pounds
- Ⓓ 4 tons

Talk MATH
Does an object that is small always weigh less than an object that is large? Explain.

Name ....

Independent Practice

Choose the best estimate for the weight of each object.

4.

- Ⓐ 1 ounce
- Ⓑ 11 ounces
- Ⓒ 1 pound
- Ⓓ 1 ton

5.

- Ⓕ 10 ounces
- Ⓖ 10 pounds
- Ⓗ 100 pounds
- Ⓘ 10 tons

6.

- Ⓐ 4 ounces
- Ⓑ 4 pounds
- Ⓒ 400 pounds
- Ⓓ 4 tons

7.

- Ⓕ 3 ounces
- Ⓖ 3 pounds
- Ⓗ 300 pounds
- Ⓘ 3 tons

8.

- Ⓐ 2 ounces
- Ⓑ 2 pounds
- Ⓒ 20 pounds
- Ⓓ 2 tons

9.

- Ⓕ 18 ounces
- Ⓖ 18 pounds
- Ⓗ 180 pounds
- Ⓘ 1 ton

10. Is it more reasonable to say that a pair of shoes weighs 1 ounce, 1 pound, or 1 ton?

    _____

11. Is it more reasonable to say that a pencil weighs 2 ounces, 2 pounds, or 2 tons?

    _____

Lesson 5 Customary Units of Weight 725

Problem Solving

12. Processes &Practices 6 Be Precise Which is most reasonable for the weight of two thoroughbred horses: 1 ounce, 1 pound, or 1 ton? Explain.

_____

_____

13. Using your answer from Exercise 12, what is a reasonable estimate for the weight of four thoroughbred horses? Explain.

_____

_____

Brain Builders

14. Estimate the weight of 4 books. Explain how you chose your unit of weight.

_____

_____

15. Processes &Practices 5 Use Math Tools Estimate the weight of three objects in your desk. Order the objects from greatest to least weight.

_____

_____

16. Building on the Essential Question How do you estimate weight? Give an example using each customary unit of weight.

_____

_____

_____

_____

Name

MY Homework

Lesson 5

Customary Units of Weight

Homework Helper

Need help? connectED.mcgraw-hill.com

Which is the best estimate for the weight of Ellie's little dog: 5 ounces, 5 pounds, or 5 tons?

You know that objects that weigh less than 1 pound are measured in ounces. It is not likely that the dog weighs less than 1 pound. Objects that are quite heavy are weighed in tons. The dog is not extremely heavy.

So, the best estimate for the weight of Ellie's dog is 5 pounds.

Practice

Choose the best estimate for the weight of each object.

1.
   - Ⓐ 6 ounces
   - Ⓑ 6 pounds
   - Ⓒ 60 pounds
   - Ⓓ 6 tons

2.
   - Ⓕ 70 pounds
   - Ⓖ 7 pounds
   - Ⓗ 70 ounces
   - Ⓘ 7 ounces

3.
   - Ⓐ 20 tons
   - Ⓑ 200 pounds
   - Ⓒ 20 pounds
   - Ⓓ 2,000 ounces

4.
   - Ⓕ 8 ounces
   - Ⓖ 80 ounces
   - Ⓗ 8 pounds
   - Ⓘ 80 pounds

Draw lines to match each object to its reasonable weight.

5. textbook • 2 ounces

6. whale • 40 tons

7. golf ball • 1 pound

8. car • 5 pounds

9. can of soup • 1.5 tons

Brain Builders

10. **Processes & Practices** **Check for Reasonableness** Jesse claims that his 3 goldfish weigh about 4 pounds together. Is Jesse's claim reasonable? Explain.

    _____
    _____

11. Allen has a 1-foot cube of foam. He claims that it weighs about 2 pounds since it is the size of two pineapples. Is Allen's claim reasonable? Explain.

    _____
    _____

Vocabulary Check

12. Write a definition for the term weight.

    _____
    _____

13. Order the units of weight from *greatest* to *least*: pound (lb), ounce (oz), ton (T).

    _____

14. **Test Practice** Which animal's weight is closest to 1 pound?

    Ⓐ an ant   Ⓒ a mosquito

    Ⓑ a lion   Ⓓ a robin

728 **Need more practice?** Download Extra Practice at connectED.mcgraw-hill.com

Name _____

Lesson 6
Convert Customary Units of Weight

ESSENTIAL QUESTION
Why do we convert measurements?

Recall that to convert from a larger unit to a smaller unit, multiply.

1 pound (lb) = 16 ounces (oz)

Multiply by 16.

Math in My World

Example 1

Leigh needs to buy 4 pounds of vegetable burger patties. The table shows the number of ounces in each package of patties. How many packages will Leigh need to buy?

Vegetable Burger Patties	
Packages	Number of Ounces
1	32
2	64
3	96
4	128

Convert 4 pounds to ounces.

Because pounds are larger than ounces, multiply.

There are 16 ounces in a pound, so multiply 4 by 16.

4 × 16 = _____

4 pounds = _____ ounces

So, Leigh will need to buy _____ package(s).

```
   □
  16
 × 4
 ──
  □□
```

Online Content at connectED.mcgraw-hill.com

A ton is a very heavy unit of measurement. There are 2,000 pounds in 1 ton.

Example 2

Use the table to find how many pounds a Stegosaurus weighed.

Dinosaur Weights	
Dinosaur	Weight (tons)
Allosaurus	2
Megalosaurus	1
Stegosaurus	3
Supersaurus	60
Tyrannosaurus	8

To find the weight of a Stegosaurus in pounds, multiply the number of tons by 2,000.

3 × 2,000 = _____

So, a Stegosaurus weighed _____ pounds.

Helpful Hint
3 × 2 = 6
3 × 20 = 60
3 × 200 = 600
So, 3 × 2,000 = 6,000.

Guided Practice

Complete each conversion table.

1.

pounds (lb)	ounces (oz)	(lb, oz)
2		(2, 32)
3		
4		
5		

Talk MATH
Explain why you multiply to convert a larger unit of measure to a smaller unit of measure.

2.

tons (T)	pounds (lb)	(T, lb)
2		
4		
6		
8		

Name ....

Independent Practice

Complete each conversion table.

3.

pounds (lb)	ounces (oz)	(lb, oz)
6		
8		
10		
12		

4.

tons (T)	pounds (lb)	(T, lb)
2		
3		
5		
7		

Algebra Find each unknown number.

5. 3 T = ■ lb

■ = _____

6. ■ oz = 2 lb

■ = _____

7. ■ lb = 6 T

■ = _____

8. $4\frac{1}{2}$ lb = ■ oz

■ = _____

9. ■ oz = 10 lb

■ = _____

10. 4 T = ■ lb

■ = _____

11. 5 lb = ■ oz

■ = _____

12. ■ lb = 3 T

■ = _____

13. ■ oz = $11\frac{1}{4}$ lb

■ = _____

14. Circle the table that represents the correct relationship between pounds and ounces.

Pounds	1	2	3	4
Ounces	16	32	48	64

Pounds	1	2	3	4
Ounces	16	24	32	40

Pounds	1	2	3	4
Ounces	8	16	24	32

Pounds	16	32	48	64
Ounces	1	2	3	4

Lesson 6 Convert Customary Units of Weight 731

Problem Solving

Baby animals weigh different amounts at birth.

Animal	Birth Weight
Alligator	2 ounces
Giraffe	100–150 pounds
Giant panda	3–5 ounces
Walrus	100–160 pounds

15. What is the least weight of a baby giraffe in ounces?

_____

16. What is the greatest weight of a baby walrus in ounces?

_____

Brain Builders

17. **Processes &Practices 4** **Model Math** If a baby walrus weighs 100 pounds, how many more ounces does it weigh than a baby alligator? Explain.

_____

_____

18. **Processes &Practices 1** **Plan Your Solution** Tiffany weighed 7 pounds 12 ounces when she was born. Her weight doubled after four months. How much did Tiffany weigh after four months? Explain.

_____

_____

_____

19. **Building on the Essential Question** How can I convert measurements of weight? Give an example.

_____

_____

_____

Name _____

Lesson 6
Convert Customary Units of Weight

Homework Helper

Need help? connectED.mcgraw-hill.com

The world's heaviest apple weighed 65 ounces. Did this apple weigh more than 4 pounds?

To answer this question, you need to convert one unit of weight to match the other.

Convert pounds to ounces.

To convert from a larger unit to a smaller unit, you need to multiply.

4 pounds × 16 ounces = 64 ounces

Since 65 > 64, the 65-ounce apple did weigh more than 4 pounds.

Use the conversion table to check.

> 65 ounces is between 4 and 5 pounds.

Helpful Hint
1 pound (lb) = 16 ounces (oz)
2,000 pounds (lb) = 1 ton (T)

pounds (lb)	ounces (oz)	(lb, oz)
4	64	(4, 64)
5	80	(5, 80)
6	96	(6, 96)
7	112	(7, 112)

Practice

Convert units to complete each equation.

1. 3 lb = _____ oz

2. 3 T = _____ lb

3. _____ oz = 10 lb

4. 9 T = _____ lb

Convert units to complete each equation.

5. $\frac{1}{2}$ ton = _____ pounds

6. $\frac{1}{4}$ pound = _____ ounces

Complete each conversion table.

7.

pounds (lb)	ounces (oz)	(lb, oz)
5		
7		
9		
11		

8.

tons (T)	pounds (lb)	(T, lb)
4		
8		
12		
16		

Problem Solving

9. Kylie bought a bunch of grapes that weighed 2 pounds and 9 ounces. How much did the grapes weigh in ounces only?

_____

10. Brent's pick-up truck can carry $\frac{1}{5}$ ton. Nina's pick-up truck can carry 500 pounds. Whose pick-up truck can carry more weight? Explain.

_____

_____

Brain Builders

11. **Processes &Practices 2** **Use Number Sense** Charlotte's family bought a 2-pound block of cheese. There is $\frac{1}{4}$ of the block of cheese left. How many ounces of cheese has Charlotte's family eaten? Explain.

_____

_____

12. **Test Practice** Which weight is greater than 5,000 pounds?

Ⓐ 3 tons Ⓑ 80 ounces Ⓒ 2,000 pounds Ⓓ 16,000 ounces

Name _____

Lesson 7
Convert Units of Time

ESSENTIAL QUESTION
Why do we convert measurements?

The steps that you use to convert units of length, weight, and capacity can be used to convert units of time.

Math in My World

Example 1

Cooper watched a butterfly in his garden for 5 minutes. How many seconds did he watch the butterfly?

1 minute = 60 seconds

To convert larger units of time to smaller units of time, multiply.

5 × 60 = _____

Helpful Hint
5 × 6 = 30
5 × 60 = 300

So, Cooper watched a butterfly for _____ seconds.

Key Concept — Units of Time

1 minute (min) = 60 **seconds** (s)
1 hour (h) = 60 min
1 day (d) = 24 h
1 week (wk) = 7 d
1 year (y) = 52 wk = 12 months (mo)

Online Content at connectED.mcgraw-hill.com

Lesson 7   735

Example 2

Audrey is going to watch a movie at the movie theater. The movie starts at 3:30 p.m. and ends at 5:37 p.m. How long is the movie in minutes?

Find the time interval, or the length of time, from the start of the movie to the end of the movie.

The time interval between 3:30 and 5:37 is _____ hours and 7 minutes.

Since 1 hour = 60 minutes, multiply 60 by 2.

60 × 2 = _____   120 + 7 = _____

So, the movie is _____ minutes long.

> Count the whole hours.
> From 3:30 to 5:30 is 2 hours.
>
> Count the remaining minutes.
> From 5:30 to 5:37 is 7 minutes.

Guided Practice

Complete each conversion table.

1.
minutes (min)	seconds (s)	(min, s)
1		(1, 60)
2		
3		
4		

Talk MATH

What operation would you use to find the number of minutes in 2 hours? Explain.

2.
hours (h)	days (d)	(h, d)
	1	
	2	
	3	
	4	

3.
years (y)	weeks (wk)	(y, wk)
1		
2		
3		
4		

736 Chapter 11 Customary Measurement

Name

Independent Practice

Complete each conversion table.

4.
days (d)	weeks (wk)	(d, wk)
	2	
	4	
	6	
	8	

5.
hours (h)	minutes (min)	(h, min)
3		
4		
5		
6		

6.
years (y)	months (mo)	(y, mo)
1		
3		
5		
7		

7.
weeks (wk)	years (y)	(wk, y)
	5	
	10	
	15	
	20	

Compare. Use <, >, or =.

8. 5 min ◯ 250 s

9. 36 mo ◯ 4 y

10. 300 s ◯ 1 h

11. How many times greater is one minute than one second? _____

12. How many times greater is one hour than one minute? _____

13. How many times greater is one week than one day? _____

14. How many minutes are in the time interval from 1:22 p.m. to 5:44 p.m.? _____

15. How many minutes are in the time interval from 7:09 a.m. to 10:36 a.m.? _____

Lesson 7 Convert Units of Time 737

Problem Solving

16. Aubrey played on the playground for $2\frac{1}{2}$ hours. How many minutes did she play?

_____

17. Alex had to rent the pavilion at the park for a minimum of 3 hours. How many minutes is that?

_____

18. Processes & Practices 5 **Use Math Tools** Justin painted his fence from 8:00 a.m. to 11:47 a.m. For how many minutes did he paint the fence?

_____

Brain Builders

19. It took Aiden 20 minutes to walk to school. It took Callie 900 seconds to walk to school. Who took less time to walk to school? Explain.

_____

_____

20. Processes & Practices 3 **Find the Error** Sadie wrote the following on the board. Find and correct her mistake.

    2 years = 24 weeks

_____

_____

21. Building on the Essential Question How does multiplication relate to time conversions? Give an example.

_____

_____

_____

738 Chapter 11 Customary Measurement

Name _____

MY Homework

Lesson 7
Convert Units of Time

Homework Helper

Need help? connectED.mcgraw-hill.com

Chloe is celebrating her 3rd birthday. How many weeks old is she?

To convert a larger unit (years) to a smaller unit (weeks), you need to multiply.

52 ← weeks in a year
× 3 ← Chloe's age in years
156

So, Chloe is 156 weeks old.

Conversion Chart Units of Time
1 minute (min) = 60 seconds (s)
1 hour (h) = 60 min
1 day (d) = 24 h
1 week (wk) = 7 d
1 year (y) = 52 wk = 12 months (mo)

Practice

Convert units to complete each equation.

1. 3 minutes = _____ seconds

2. 5 weeks = _____ days

3. _____ months = 5 years

4. _____ minutes = 6 hours

5. 4 days = _____ hours

6. _____ weeks = 8 years

7. $1\frac{1}{2}$ days = _____ hours

8. $3\frac{2}{7}$ weeks = _____ days

9. _____ months = $4\frac{3}{4}$ years

10. _____ minutes = 6 hours and 42 minutes

Complete each conversion table.

11.
weeks (wk)	days (d)	(wk, d)
2		
4		
6		
8		

12.
minutes (min)	hours (h)	(min, h)
	9	
	7	
	5	
	3	

Problem Solving

13. Emma is $9\frac{1}{4}$ years old. How many months old is Emma?

    _____

14. **Processes & Practices 4** **Model Math** Vincent is watching a movie that lasts 1 hour and 37 minutes. He has watched 52 minutes so far. How many minutes are left in the movie?

    _____

Brain Builders

15. Lexie started her homework at 4:30 p.m. She finished at 5:05 p.m. How many seconds did it take her to finish her homework? Explain.

    _____

    _____

Vocabulary Check

16. How many seconds are in 1 minute? _____

17. **Test Practice** Cameron's book log shows that he read a total of $1\frac{1}{2}$ hours each week for 4 weeks. How many total minutes did he read?

    Ⓐ 60 minutes   Ⓒ 360 minutes

    Ⓑ 90 minutes   Ⓓ 480 minutes

740 **Need more practice?** Download Extra Practice at connectED.mcgraw-hill.com

Check My Progress

Vocabulary Check

1. Choose a vocabulary word to complete each sentence.

   ounce   **pound**   **second**
   ton   **weight**

   - A rhinoceros weighs about 2 _____ .
   - A mouse weighs about 1 _____ .
   - A squirrel weighs about 3 _____ .
   - The _____ of an object describes how heavy it is.
   - There are 60 _____ in a minute.

Concept Check

Complete each conversion table.

2.
ounces (oz)	pounds (lb)	(oz, lb)
	1	
	3	
	5	
	7	

3.
hours (h)	minutes (min)	(h, min)
2		
4		
6		
8		

Check My Progress 741

Problem Solving

4. Is it more reasonable to say that a pen weighs 2 ounces, 2 pounds, or 2 tons?

5. A hippopotamus eats 100 pounds of food a day. How many days would it take the hippo to eat 1 ton of food?

Brain Builders

6. An ostrich egg weighs 64 ounces. Is the weight of the ostrich egg greater than 5 pounds? Explain.

7. Jordan spent 3 hours and 12 minutes at a local farm. How many minutes did he spend at the farm?

8. **Test Practice** Which is the most reasonable estimate for the weight of a bowling ball?

Ⓐ 8 tons
Ⓑ 80 pounds
Ⓒ 8 pounds
Ⓓ 8 ounces

742 Chapter 11 Customary Measurement

Name _____

Lesson 8
Display Measurement Data in a Line Plot

ESSENTIAL QUESTION
Why do we convert measurements?

You can represent measurement data for fractions of a unit in a **line plot**. The line plot's number line will look like a ruler.

Math in My World

Example 1

The science room has a collection of bugs. Each bug's length is measured to the nearest eighth of an inch. Make a line plot to represent the data.

The measurement data are shown in the tally chart.

 First, make a number line to represent the value of each bug's length.

Bug Length	
$\frac{1}{8}$ in.	
$\frac{2}{8}$ in.	||
$\frac{3}{8}$ in.		
$\frac{4}{8}$ in.	||
$\frac{5}{8}$ in.	|||||
$\frac{6}{8}$ in.	|||||
$\frac{7}{8}$ in.	||||
$\frac{8}{8}$ in.		

 Next, place an X above each measurement every time that value occurred.

Online Content at connectED.mcgraw-hill.com

Lesson 8 743

Example 2

Refer to Example 1. Find the difference in length between the longest and shortest bug.

To find the difference between the longest and shortest bugs, subtract the shortest length from the longest length.

Subtract like fractions.

$$\frac{8}{8} - \frac{2}{8} = \frac{8-2}{8} = \frac{6}{8}$$

$$= \frac{3}{4}$$

So, the difference between the longest and the shortest bugs in the collection is _____ of an inch.

Guided Practice

For Exercises 1–2, use the tally chart shown.

Button Sizes	
$\frac{1}{8}$ in.	ⵏⵏⵏⵏ I
$\frac{3}{8}$ in.	III
$\frac{4}{8}$ in.	I
$\frac{5}{8}$ in.	II
$\frac{7}{8}$ in.	IIII

1. The tally chart represents the widths of buttons collected by Bella's mother. Represent this data in a line plot.

Button Sizes (in.)

2. Suppose the buttons that were $\frac{3}{8}$-inch wide were laid in a row touching. How far would that row extend?

_____

Talk MATH
Describe a real-world situation in which the data in a tally chart and line plot could be helpful.

744 Chapter 11 Customary Measurement

Name _____

Independent Practice

For Exercises 3–6, use the table shown.

3. **Processes & Practices 4** **Model Math** The frequency table represents fractions of an hour Sonja studied each evening over the last twelve days. Represent this data in a line plot.

Study Time			
$\frac{1}{4}$ h	$\frac{3}{4}$ h	$\frac{1}{2}$ h	$\frac{1}{4}$ h
$\frac{1}{2}$ h	$\frac{3}{4}$ h	$\frac{1}{4}$ h	1 h
$\frac{1}{2}$ h	$\frac{1}{4}$ h	$\frac{3}{4}$ h	$\frac{1}{2}$ h

Study Time (hr)

4. Which time interval was least frequent as a study time for Sonja? Explain.

_____

_____

5. What is the difference in the total time spent studying during the $\frac{3}{4}$-hour time intervals and the $\frac{1}{2}$-hour intervals?

_____

6. What is the total time Sonja spent studying over the last twelve days in minutes? What is the equivalent time in hours and minutes?

_____

Lesson 8 Display Measurement Data in a Line Plot **745**

Problem Solving

For Exercises 7–9, use the table shown.

A Week of Water for Quinn's Pets			
hamster $\frac{1}{8}$ c	dog 1c	hamster $\frac{2}{8}$ c	rabbit $\frac{1}{8}$ c
rabbit $\frac{2}{8}$ c	hamster $\frac{1}{8}$ c	cat $\frac{4}{8}$ c	dog 1c
cat $\frac{5}{8}$ c	dog 1c	rabbit $\frac{1}{8}$ c	dog $\frac{6}{8}$ c

7. Every time one of Quinn's pets needed its water refilled over the past week, he marked the amount of water given in a table. Represent the data in a line plot.

8. What is the difference between the smallest amount of water and the greatest amount of water Quinn gave to his pets over the week? Explain.

Brain Builders

9. **Processes & Practices 3** **Draw a Conclusion** How much more water did the cat get than the hamster over the course of the week? Explain.

10. **Processes & Practices 6** **Explain to a Friend** Explain to a friend the similarity between the number lines on a line plot and the markings on a ruler. Draw models to illustrate your answer.

11. **Building on the Essential Question** Line plots can be used to display measurement data. Name another way to display measurement data.

746 Chapter 11 Customary Measurement

Name _____

MY Homework

Lesson 8

Display Measurement Data in a Line Plot

Homework Helper

Need help? connectED.mcgraw-hill.com

At the farmers' market, Jeannie sells beans. Her scale weighs them in fractions of pounds. Jeannie made a tally chart showing how much each handful of beans weighed. How much do the beans weigh all together?

Weight of Green Beans	
$\frac{1}{4}$ pound	||||| ||
$\frac{1}{2}$ pound	|||||	
$\frac{3}{4}$ pound	||||
1 pound	||

1 Make a line plot to represent the data. First, draw a number line. Then place an X above each measurement for each time that weight occurred.

2 How much do the beans weigh all together? Multiply to find the total weight for each value.

$7 \times \frac{1}{4}$ lb = $\frac{7}{4}$ lb = $1\frac{3}{4}$ lb

$6 \times \frac{1}{2}$ lb = $\frac{6}{2}$ lb = 3 lb

$4 \times \frac{3}{4}$ lb = $\frac{12}{4}$ lb = 3 lb

2×1 lb = 2 lb

Then add the weights to find the total.

$1\frac{3}{4}$ lb + 3 lb + 3 lb + 2 lb = $9\frac{3}{4}$ lb

So, all together the beans weigh $9\frac{3}{4}$ pounds.

Weight of Green Beans (lb)

Practice

1. Refer to the Homework Helper. What is the difference between the greatest green bean weight and the least green bean weight?

_____

For Exercises 2 and 3, use the tally chart shown. The tally chart represents the distance some children were able to ride on a unicycle.

Distance	
$\frac{1}{5}$ mi	|||||	
$\frac{2}{5}$ mi	||||
$\frac{3}{5}$ mi	||
$\frac{4}{5}$ mi	||
1 mi		

2. Represent this data in a line plot.

3. What is the difference between the greatest distance ridden and the least distance ridden?

Brain Builders

4. **Processes & Practices 3** **Draw a Conclusion** Walter practices piano for either $\frac{1}{4}$ hour, $\frac{1}{2}$ hour, or $\frac{3}{4}$ hour each day. If a line plot shows two Xs above each value of time, what is the total amount of time Walter has spent practicing? Explain.

Vocabulary Check

5. Describe one way to use a line plot.

6. **Test Practice** Look at the tally chart or line plot from Exercise 2 above. What was the total distance ridden by all of the children?

Ⓐ $5\frac{3}{5}$ miles

Ⓒ $6\frac{2}{5}$ miles

Ⓑ $5\frac{4}{5}$ miles

Ⓓ $6\frac{3}{5}$ miles

748 Need more practice? Download Extra Practice at connectED.mcgraw-hill.com

Name _____

Lesson 9
Solve Measurement Problems

ESSENTIAL QUESTION
Why do we convert measurements?

Math in My World

Example 1

The blue trail at the park is $\frac{3}{4}$-mile long. The red trail is $1\frac{1}{4}$-miles long. Stella wants to hike both trails. How many feet will she hike in all?

 Find the total number of miles.

$$1\frac{1}{4} + \frac{3}{4} = \boxed{\frac{4}{4}} + \frac{1}{4} + \frac{3}{4}$$

$$= \frac{5}{4} + \frac{3}{4}$$

$$= \frac{8}{4}$$

$$= 2$$

The number line shows that $\frac{3}{4} + \frac{5}{4} = \frac{8}{4}$, or 2.

Stella will hike 2 miles.

 Convert the miles to feet.

1 mile = 5,280 feet

Multiply 2 miles by 5,280 to find the total number of feet.

So, Stella will hike _____ feet in all.

Online Content at connectED.mcgraw-hill.com

Lesson 9 749

Example 2

Dominic weighed an apple and a watermelon. The apple weighed 5 ounces. The watermelon weighed 20 pounds. How many more ounces did the watermelon weigh than the apple?

  Find the total number of ounces the watermelon weighed.

There are 16 ounces in a pound. Multiply 20 by 16.

```
   20
 × 16
  120
+ 200
  320
```

The watermelon weighed 320 ounces.

  Find the difference in ounces.

320 oz − 5 oz = _____ oz

So, the watermelon weighed _____ ounces more than the apple.

Guided Practice

1. Stan has 4 pints of milk. Gianna has twice as much milk as Stan. How many fluid ounces of milk does Gianna have?

    4 pints × 2 = _____ pints

    Gianna has _____ pints of milk.

    8 pints × 16 = _____ fluid ounces

    Gianna has _____ fluid ounces of milk.

Which operations did you use to solve Exercise 2? Explain.

2. Molly went to her grandma's house for 16 hours last weekend. This was four times longer than the time she spent at her grandma's this weekend. How many minutes was Molly at her grandma's house this weekend?

    _____

Name _____

Independent Practice

3. Jayden read to his little sister for 10 minutes on Saturday and 12 minutes on Sunday. How many seconds did he read to her on Saturday and Sunday? Circle what to do to solve this problem.

   - Add 10 and 12 and then multiply by 60.
   - Add 10 and 12 and then multiply by 30.
   - Add 10 and 12 and then divide by 60.
   - Add 10 and 12 and then subtract 60.

   Solve.

   Jayden read to his little sister for _____.

4. Mackenzie's dog can jump $2\frac{1}{2}$ feet off the ground. Jordyn's dog can jump $\frac{1}{2}$ foot off the ground. What is the difference in inches between how high Mackenzie's dog can jump and how high Jordyn's dog can jump? Circle what to do to solve this problem.

   - Subtract $2\frac{1}{2}$ ft $-\frac{1}{2}$ ft and then multiply by 3.
   - Subtract $2\frac{1}{2}$ ft $-\frac{1}{2}$ ft and then divide by 3.
   - Subtract $2\frac{1}{2}$ ft $-\frac{1}{2}$ ft and then multiply by 12.
   - Add $2\frac{1}{2}$ ft $+\frac{1}{2}$ ft and then multiply by 12.

   Use the number line to help solve.

   The difference in inches is _____.

Lesson 9 Solve Measurement Problems 751

Problem Solving

5. Brooklyn bought 1 pound of cucumbers for a salad. She bought twice as much lettuce. How many ounces of lettuce did Brooklyn buy for the salad?

   _____

6. **Processes & Practices** 5 **Use Math Tools** Sebastian needs 2 gallons of juice for his party. How many cups of juice is this?

   _____

7. The practice field is 13 yards long. How many feet is this?

   _____

Brain Builders

8. Jacob needs 1 pint of milk for his recipe. He needs $1\frac{1}{2}$ cups of water. How many cups will he have after he combines both ingredients? Explain.

   _____

   _____

9. **Processes & Practices** 2 **Use Number Sense** The moving company charges extra money to move boxes weighing more than 25 pounds. Circle the packages that weigh less than 25 pounds. Put an X on the packages that weigh more than 25 pounds.

   320 ounces   480 ounces   288 ounces

   368 ounces   352 ounces   432 ounces

10. **Building on the Essential Question** What information do I need to solve word problems that involve measurement and why do I need it?

    _____

    _____

752 **Chapter 11** Customary Measurement

Name _____

MY Homework

Lesson 9

Solve Measurement Problems

Homework Helper

Need help? connectED.mcgraw-hill.com

Harlan walked his dog $\frac{1}{6}$ mile on Monday, $\frac{3}{6}$ mile on Tuesday, and $\frac{5}{6}$ mile on Wednesday. What is the total distance Harlan walked his dog that week?

Add the three distances. $\quad \frac{1}{6} + \frac{3}{6} + \frac{5}{6} = \frac{9}{6}$

Simplify. $\quad \frac{9}{6} = 1\frac{3}{6}$ or $1\frac{1}{2}$

So, Harlan walked his dog a total of $1\frac{1}{2}$ miles.

Practice

For Exercises 1–3, use the following information.
Marissa bought $2\frac{1}{2}$ gallons of orange juice to make punch.

1. How many quarts of orange juice did Marissa buy?

   _____

2. Marissa bought 6 fewer pints of soda than of orange juice. How many cups of soda did she buy?

   _____

3. Marissa pours $1\frac{1}{2}$ cups of punch in each glass. How many fluid ounces is each serving?

   _____

Lesson 9 My Homework 753

Problem Solving

For Exercises 4–6, use the following information.
Martin sorted the bolts in his toolbox by size. They measured $\frac{3}{8}$ inch, $\frac{1}{2}$ inch, $\frac{5}{8}$ inch, $\frac{7}{8}$ inch, and $1\frac{1}{4}$ inches.

4. What is the difference in length between the longest bolt and the shortest bolt?

   _____

5. If you laid one of each size bolt end to end, how long would the row of bolts be?

   _____

Brain Builders

6. If you laid 16 bolts that each measure $\frac{7}{8}$ inch end to end, how many feet long would the row of bolts be?

   _____

7. Aurora has lived in her current house for 6 years. Mitch has lived in his current house for 1,460 days. Which student has lived in their current house longer? How much longer?

   _____

8. **Processes & Practices 5** **Use Math Tools** Nolan bought 26 yards of heavy-duty rope at the hardware store. The rope costs $2 per foot. How much did Nolan pay for the rope? Explain.

   _____

9. **Test Practice** Kim and her two brothers each use $1\frac{1}{2}$ cups of milk for breakfast. How many fluid ounces of milk do they use in 4 days?

   Ⓐ 144 fl oz   Ⓒ 72 fl oz

   Ⓑ 108 fl oz   Ⓓ 36 fl oz

Name _____

Lesson 10
Problem-Solving Investigation
STRATEGY: Guess, Check, and Revise

ESSENTIAL QUESTION
Why do we convert measurements?

Learn the Strategy

Three elk are walking in the mountains. Two of the elk weigh the same amount. The other elk weighs 150 pounds more than the other two. If the total weight of the three elk is 1,650 pounds, how much does each elk weigh?

1 Understand

What facts do you know?
Two elk weigh the same amount. A third elk weighs 150 pounds more than the other two.

The total weight of the three elk is _____ pounds.

What do you need to find?
the weight of each elk

2 Plan

I will guess, check, and revise to solve the problem.

3 Solve

First Elk Weight (lb)	Second Elk Weight (lb)	Third Elk Weight (lb)	Total Weight (lb)	Check
400	400	550	400 + 400 + 550 = 1,350	too low
500	500	650	500 + 500 + 650 = 1,650	correct

So, two of the elk weigh _____ pounds. The third elk weighs _____ + 150, or _____ pounds.

4 Check

Does your answer make sense? Explain.

_____

Online Content at connectED.mcgraw-hill.com

Lesson 10 755

Practice the Strategy

Corrinne is making twice as much fruit punch as lemonade. She is making 12 gallons total. How many gallons will be fruit punch and how many will be lemonade?

 Understand

What facts do you know?

___

___

What do you need to find?

___

 Plan

___

 Solve

 Check

Does your answer make sense? Explain.

___

___

756 Chapter 11 Customary Measurement

Name _____

Apply the Strategy

Guess, check, and revise to solve each problem.

1. **Processes & Practices 3** **Draw a Conclusion** Theo lives twice as far from Cassidy as Jarvis. How far do Theo and Jarvis each live from Cassidy?

_____

2. Jorge and his two brothers each water the plants. Jorge's watering can holds half as much water as his older brother's watering can. His younger brother's watering can holds 8 cups. All together, the watering cans hold 2 gallons. How many cups does Jorge's watering can hold?

_____

Brain Builders

3. **Processes & Practices 1** **Make Sense of Problems** Rhonda filled her aquarium with water. Then, she took out 3 gallons to make room for the gravel and fish. She added 2 quarts back in. Then, she added 4 fish. Each fish was in a separate bag with 1 pint of water. If she has a 20-gallon aquarium, how many pints are in the aquarium now? Explain.

_____

_____

Lesson 10 Problem-Solving Investigation

Review the Strategies

Use any strategy to solve each problem.
- Guess, check, and revise.
- Find extra or missing information.
- Use logical reasoning.
- Look for a pattern.

4. The distance from Ian's home to the museum is 2,640 yards. Is it reasonable to say that Ian's home is more than 9,000 feet away from the museum? (*Hint:* 3 feet = 1 yard)

___

___

___

5. Processes & Practices 5 **Use Math Tools** One seal weighs 22 pounds. Another seal weighs three times as much. How much do the seals weigh all together?

___

6. Processes & Practices 4 **Model Math** Mario wants to download 12 songs on his digital music player. He has only 5 minutes to download the songs. If it takes 30 seconds for Mario to download one song, will he have enough time to download all of the songs? Explain.

___

___

7. A stunt person jumps from the roof of a 54-foot building. A skydiver jumps from a plane that is 186 times as high. From what height did the skydiver jump?

___

758 Chapter 11 Customary Measurement

Name

MY Homework

Lesson 10

Problem Solving: Guess, Check, and Revise

Homework Helper

Need help? connectED.mcgraw-hill.com

Cassie is doing sand art. She fills a bottle with 10 inches of sand. She makes two equal layers of yellow sand with a layer of blue sand in between. The layer of blue sand is 1 inch greater than each layer of yellow sand. How many inches of each color of sand does she use?

1 Understand

What facts do you know?

The bottle holds 10 inches of sand. There are two equal layers of yellow sand. There is one layer of blue sand that is 1 inch greater than each yellow layer.

What do you need to find?

the number of inches of each color of sand used

2 Plan

Guess, check, and revise to solve the problem.

3 Solve

Layers of Sand (in.)			Total (in.)	Check
1st Yellow Layer (in.)	2nd Yellow Layer (in.)	Blue Layer (in.)		
4	4	5	4 + 4 + 5 = 13	too high
3	3	4	3 + 3 + 4 = 10	correct

So, there are two 3-inch layers of yellow sand with a 4-inch layer of blue sand in between the yellow sand layers.

4 Check

Does the answer make sense?

Yes; 3 + 3 + 4 = 10 and 4 − 3 = 1.

Problem Solving

Guess, check, and revise to solve each problem.

1. Max was on vacation twice as long as Jared and half as long as Wesley. The boys were on vacation a total of 3 weeks. How many days was each boy on vacation?

2. Anu drinks 2 cups of water each day. Jan drinks twice as much water as Anu. How many fluid ounces does Jan drink each day?

Brain Builders

3. **Processes & Practices** **Plan Your Solution** Casey likes to run. She runs an additional $\frac{1}{4}$ mile each day. On the last day, she ran $1\frac{1}{4}$ miles. If she ran $\frac{1}{2}$ mile her first day, for how many days has she been running? Explain.

4. There are 4 semi-trailer trucks parked in a line at the rest stop. After the first truck, each truck in the line weighs 2 tons more than the truck before it. The trucks weigh a total of 32 tons. How many pounds does each truck weigh? Explain.

Review

Chapter 11
Customary Measurement

Vocabulary Check

Draw a line from each word to its definition.

1. customary system
2. convert
3. line plot
4. capacity
5. weight

- A measurement that tells the amount of liquid a container can hold.
- A measurement that tells how heavy an object is.
- A graph that uses columns of Xs above a number line to show frequency of data.
- To change one unit to another.
- Units of measure most often used in the United States.

Write an example for each measurement.

6. 1 yard _____
7. 1 quart _____
8. 1 foot _____
9. 1 gallon _____
10. 1 mile _____
11. 1 ounce _____
12. 1 fluid ounce _____
13. 1 pound _____
14. 1 cup _____
15. 1 ton _____
16. 1 pint _____
17. 1 second _____

My Chapter Review 761

Concept Check

Complete each conversion table.

18.

inches (in.)	yards (yd)	(in., yd)
	1	
	3	
	5	
	7	

19.

cups (c)	gallons (gal)	(c, gal)
	2	
	3	
	4	
	5	

20.

feet (ft)	yards (yd)	(ft, yd)
	2	
	4	
	6	
	8	

21.

minutes (min)	seconds (s)	(min, s)
1		
3		
5		
7		

22. How many times greater is one hour than one minute? _____

23. How many times greater is one gallon than one quart? _____

24. How many times greater is one mile than one foot? _____

25. How many times greater is one pound than one ounce? _____

26. How many minutes have passed between the time shown on the first clock and the time shown on the second clock?

_____

27. How many seconds have passed between the time shown on the first clock and the time shown on the second clock?

_____

Problem Solving

For Exercises 28–31, use the table shown. The table represents the lengths of beads that Zoe is using to make a necklace.

Lengths of Beads		
$\frac{1}{6}$ in.	$\frac{4}{6}$ in.	$\frac{5}{6}$ in.
$\frac{2}{6}$ in.	$\frac{1}{6}$ in.	$\frac{5}{6}$ in.
$\frac{5}{6}$ in.	$\frac{1}{6}$ in.	$\frac{4}{6}$ in.

28. Represent this data in a line plot.

Brain Builders

29. What is the total length of the beads that are longer than $\frac{1}{2}$-inch long?

_____

30. How many more of the $\frac{1}{6}$-inch long beads would Zoe need if she wants to make a 1-foot long necklace out of only $\frac{1}{6}$-inch long beads?

_____

31. Test Practice Which bead length has enough beads to total exactly $\frac{1}{2}$ inch if the beads of that size are laid end to end?

Ⓐ $\frac{1}{6}$ inch Ⓒ $\frac{4}{6}$ inch

Ⓑ $\frac{2}{6}$ inch Ⓓ $\frac{5}{6}$ inch

Reflect

Chapter 11
Answering the
ESSENTIAL QUESTION

Use what you learned about customary measurements to complete the graphic organizer.

Length

Time

Weight

ESSENTIAL QUESTION
Why do we convert measurements?

Capacity

Reflect on the **ESSENTIAL QUESTION** Write your answer below.

_____

_____

_____

764 Chapter 11 Customary Measurement

Name _____ Date _____

Score _____

Performance Task

Brain Builders

Darrel Delivers!

Darrel drives a delivery truck. He has to be a very careful driver as well as a very fit and healthy person to carry all the boxes he delivers.

Show all your work to receive full credit.

Part A

Darrel has a large box of perfume on his truck for delivery to a department store. The box holds 32 bottles that each contain 8 fluid ounces of perfume. How many pints of perfume are in the box? Show your work.

Part B

Darrel works out on a treadmill. So far he has recorded the following distances:

Darrel's Distances (in miles)			
$\frac{3}{8}$	$\frac{5}{8}$	$\frac{3}{8}$	$\frac{4}{8}$
$\frac{7}{8}$	$\frac{2}{8}$	$\frac{6}{8}$	$\frac{7}{8}$
$\frac{5}{8}$	$\frac{6}{8}$	$\frac{5}{8}$	$\frac{4}{8}$
$\frac{3}{8}$	$\frac{6}{8}$	$\frac{5}{8}$	$\frac{7}{8}$

Complete the line plot using the information in the table. For how many days did Darrel run $\frac{5}{8}$ mile? Was this more or fewer days than he ran $\frac{7}{8}$ mile?

Online Content at connectED.mcgraw-hill.com

Performance Task   764PT1

Part C

Darrel is delivering appliances today. He has 3 washers, 2 microwaves, 4 dishwashers, and 6 grills on the truck. The table shows the weight of one of each appliance. Does Darrel have more than 1 ton on his truck? If so, how much more?

Appliance	Weight (in pounds)
washer	205
microwave	60
dishwasher	110
grill	150

Part D

Darrel delivered furniture and other household items to a family that had moved from another city. He made 9 round trips walking from the truck to the side door and back, a distance of 46 yards. He made 21 trips from the truck to the front door and back, a distance of 54 yards. He made 5 trips from the truck to the garage and back, a distance of 63 yards. Did Darrel walk more than or less than 1 mile in all? (Hint: There are 5,280 feet in 1 mile.)

Part E

Darrel created a conversion table for the length of his breaks during the day. Complete the table. Darrel's supervisor told him to take a 1,320-second break. What is that time in minutes?

seconds (s)	minutes (min)	(s, min)
60	1	(60, 1)
300		
600		
	18	
	27	

764PT2 **Chapter 11** Customary Measurement

Chapter 12 Metric Measurement

ESSENTIAL QUESTION

How can conversion of measurements help me solve real-world problems?

Around My House

Watch a video!

Name _____

MY Chapter Project

Metric Units of Capacity

1. Look through your grocery ad or newspaper to find products sold in milliliter and liter containers. Cut out pictures of the products and write the products and their measures in the tables below.

Milliliters

Product	Measure in milliliters

Liters

Product	Measure in liters

2. Glue the pictures on a poster board. Glue the products sold in milliliter containers onto the left side of the poster board, and the products sold in liter containers onto the right side.

3. Think of a question you could ask the class about your poster.

Question: _____

_____

Name _____

Am I Ready?

Multiply.

1. 8 × 10 = _____
2. 7 × 1,000 = _____
3. 10 × 3 = _____
4. 4 × 1,000 = _____
5. 10 × 9 = _____
6. 1,000 × 6 = _____

7. Ted is comparing his shoe to his mom's shoe. Circle the longer shoe.

Ted's shoe   Mom's shoe

Write each fraction as a decimal.

8. $\frac{2}{10}$ = _____
9. $\frac{6}{10}$ = _____
10. $\frac{9}{10}$ = _____
11. $\frac{24}{100}$ = _____
12. $\frac{16}{100}$ = _____
13. $\frac{83}{100}$ = _____

Write a decimal for each part of a dollar shown.

14.

15.

How Did I Do? Shade the boxes to show the problems you answered correctly.

1 2 3 4 5 6 7 8 9 10 11 12 13 14 15

Name ....

MY Math Words

Review Vocabulary

capacity   length

Making Connections

Define and provide examples for each review vocabulary word.

	Capacity	Length
Define.		
Provide examples.		

My Vocabulary Cards

Lesson 12–1

centimeter (cm)

Lesson 12–3

gram (g)

5 grams

Lesson 12–3

kilogram (kg)

1 kilogram

Lesson 12–1

kilometer (km)

6 city blocks = 1 kilometer

Lesson 12–2

liter (L)

1 liter   about 5 liters

Lesson 12–3

mass

less mass   more mass

Lesson 12–1

meter (m)

about 1 meter

Lesson 12–1

metric system

1 gram   1 liter

Ideas for Use

- Draw or write examples for each card. Be sure your examples are different from what is shown on each card.
- Work with a partner to name the part of speech of each word. Consult a dictionary to check your answers.

A metric unit for measuring mass.

Name another unit of measurement that contains the word part *gram*.

_____

_____

A metric unit for measuring length.
100 centimeters = 1 meter

The prefix *cent-* means "hundred." Write another math word with this word part and its meaning.

_____

_____

A metric unit for measuring length.
1,000 meters = 1 kilometer

Examine the conversions for this card and *kilogram*. Identify and define the prefix. Explain how it can help you recall the words.

_____

_____

A metric unit for measuring mass.
1,000 grams = 1 kilogram

Explain whether or not you would weigh a feather in kilograms.

_____

_____

The amount of matter in an object.

Name two objects. Which has a greater mass?

_____

_____

A metric unit for measuring volume or capacity. 1 liter = 1,000 milliliters

Would you use liters or milliliters to measure the capacity of a bottle of orange juice?

_____

_____

A system of measurement that includes meters, grams, and liters.

Compare and contrast a unit in the customary system with a unit in the metric system.

_____

_____

A metric unit for measuring length.

Estimate the distance from your desk to the classroom door in meters.

_____

_____

MY Vocabulary Cards

Lesson 12–2

milliliter (mL)

250 mL

Lesson 12–1

millimeter (mm)

Ideas for Use

- Group 2 or 3 common words. Add a word that is unrelated to the group. Then work with a friend to name the unrelated word.

- Write the name of each lesson on the front of each blank card. Write a few study tips for each lesson on the back of each card.

A metric unit for measuring length.
1 centimeter = 10 millimeters

Would you use millimeters to measure the length of your classroom? Explain.
_____
_____

A metric unit for measuring capacity.
1,000 milliliters = 1 liter

Name two items in your home that you could measure in milliliters.
_____
_____

MY Foldable

FOLDABLES® Follow the steps on the back to make your Foldable.

Metric Conversions

Length
1 centimeter (cm) = 10 millimeters (mm)
1 meter (m) = 100 centimeters (cm)
1 kilometer (km) = 1,000 meters (m)

Capacity
1 liter (L) = 1,000 milliliters (mL)

Mass
1 kilogram (kg) = 1,000 grams (g)

Metric Measurements

Mass

gram (g)

meter (m)

kilometer (km)

FOLDABLES Study Organizer

Length

millimeter (mm)

kilogram (kg)

centimeter (cm)

Capacity

milliliter (mL)

liter (L)

Name ....

Lesson 1
Metric Units of Length

ESSENTIAL QUESTION

How can conversion of measurements help me solve real-world problems?

Length is the measurement of a line between two points. Millimeter, centimeter, meter, and kilometer are units that are part of the **metric system** of measure for length.

A **millimeter (mm)** is about as thick as 6 sheets of notebook paper.

A **centimeter (cm)** is about the length of a ladybug.

A **meter (m)** is about the height of a chair.

A **kilometer (km)** is about six city blocks.

Math in My World

Example 1

Doug is growing carrots in his garden. He pulled out a carrot for lunch. Measure the carrot to the nearest centimeter.

Align the 0 on the ruler with the left side of the carrot.

The carrot is closer to the 12-centimeter mark than the 11-centimeter mark.

So, the carrot is almost _____ centimeters long.

Online Content at connectED.mcgraw-hill.com

Lesson 1 775

Before measuring the length of an object, always estimate the length to decide which unit of measurement is best to use.

Example 2

Choose the best estimate for the length of a student's desk.

10 millimeters = 1 centimeter

Ⓐ 5 centimeters

Ⓑ 5 millimeters

Ⓒ 50 centimeters

Ⓓ 50 millimeters

A desk has to be long enough to work on. So, centimeters are a better estimate than millimeters.

Since 5 centimeters, 5 millimeters, and 50 millimeters are all too small, the answer is _____, or choice ____.

Guided Practice

1. Estimate the length of the grasshopper. Then measure to the nearest centimeter.

2. Choose the best estimate for the length of the kayak.

   Ⓐ 6 centimeters

   Ⓑ 2 meters

   Ⓒ 6 meters

   Ⓓ 2 kilometers

Talk MATH

Describe a situation when it would be appropriate to measure an object using millimeters.

776 **Chapter 12** Metric Measurement

Name

Independent Practice

Estimate each length. Then measure each object to the nearest centimeter.

3.

Length: _____

4.

Length: _____

5.

Length: _____

6.

Length: _____

Choose the best estimate for each length.

7. height of a cornstalk

    Ⓐ 2 millimeters
    Ⓑ 2 centimeters
    Ⓒ 2 meters
    Ⓓ 2 kilometers

8. length of an airport runway

    Ⓕ 5 millimeters
    Ⓖ 50 centimeters
    Ⓗ 5 meters
    Ⓘ 5 kilometers

Lesson 1 Metric Units of Length 777

Problem Solving

9. A giraffe at the zoo is 5 meters tall. Name something else that is about 5 meters tall.

_____

10. Is the distance from Boston, Massachusetts, to Phoenix, Arizona, about 4,000 centimeters or 4,000 kilometers? Explain.

_____

Brain Builders

11. Processes & Practices 3 Justify Conclusions Why would it be better to measure the length of your classroom with a meterstick instead of a centimeter ruler?

_____

_____

Estimate the length of your classroom.

_____

12. Processes & Practices 5 Use Math Tools Find three things in the classroom that are longer than 10 centimeters and smaller than 100 centimeters. Estimate the length of each item.

_____

_____

13. Building on the Essential Question Is it reasonable to use centimeters to measure the length of any object? Explain.

_____

_____

Name ....

MY Homework

Lesson 1
Metric Units of Length

Homework Helper

Need help? connectED.mcgraw-hill.com

Estimate the length. Then measure to the nearest centimeter.

1 Estimate.
You know that the length of a ladybug is about 1 centimeter. You can estimate that the pencil is about 10 ladybugs, or 10 centimeters, long.

2 Measure.
Use a centimeter ruler. Line up the 0 on the ruler with the end of the pencil. The pencil ends just after the 11-centimeter mark.

So, the pencil is about 11 centimeters long.

Practice

Estimate each length. Then measure each object to the nearest centimeter.

1.   Length: _____

2.   Length: _____

3.   Length: _____

Choose the best estimate for each length.

4. length of a river
   - Ⓐ 27 km
   - Ⓒ 170 cm
   - Ⓑ 7 m
   - Ⓓ 270 mm

5. length of a sunflower seed
   - Ⓕ 90 cm
   - Ⓗ 90 mm
   - Ⓖ 9 cm
   - Ⓘ 9 mm

Problem Solving

6. **Processes & Practices 3** **Draw a Conclusion** Sonia is standing 20 centimeters from the door. Brice is standing 20 meters from the door. Who is standing farther from the door?

   _____

7. Carly says she walks 300 millimeters to school each day. Is this reasonable? Explain.

   _____

Brain Builders

8. At his aunt's farm, Benjamin sees a horse and estimates that it is about 2 centimeters long. Is Benjamin's estimate reasonable? Explain.

   _____

Vocabulary Check

9. List the metric system units for measuring length in order from *greatest* to *least*.

   centimeter   kilometer   meter   millimeter

   _____

10. **Test Practice** Which is the best unit to use for measuring the length of an eyelash?
    - Ⓐ millimeter
    - Ⓑ centimeter
    - Ⓒ meter
    - Ⓓ kilometer

Name ....

Lesson 2
Metric Units of Capacity

ESSENTIAL QUESTION
How can conversion of measurements help me solve real-world problems?

The amount of liquid a container can hold is its capacity. The liter (L) and milliliter (mL) are units of measurement for capacity in the metric system.

liter (L)

A bottle this size can hold a liter.

milliliter (mL)

An eyedropper holds about one milliliter.

Math in My World

Example 1

Decide whether 300 milliliters or 300 liters is the more reasonable estimate for the capacity of the mug.

Use logic to estimate the capacity.

300 mL

Helpful Hint
300 eyedroppers are reasonable.

300 L

Helpful Hint
300 bottles are too much.

So, _____ is the more reasonable estimate.

Online Content at connectED.mcgraw-hill.com

Lesson 2 781

Example 2

Decide whether 600 milliliters or 600 liters is the more reasonable estimate for the capacity of the swimming pool.

The pool is a large object. So, 600 milliliters is too small.

So, _____ is the more reasonable estimate.

Guided Practice

Circle the more reasonable estimate for each capacity.

1.

    1 mL

    1 L

2.

    38 mL

    38 L

3.

    220 mL

    220 L

Talk MATH

Describe the unit of capacity you would use to measure the capacity of a bottle of medicine.

782 Chapter 12 Metric Measurement

Name ....

Independent Practice

Circle the more reasonable estimate for each capacity.

4.

   150 mL

   150 L

5.

   120 mL

   120 L

6.

   500 mL

   500 L

7.

   700 mL

   700 L

8.

   1 mL

   1 L

9.

   30 mL

   30 L

10. Select three containers. Decide whether each container has a capacity that is greater than, less than, or equal to 1 liter. Complete the table.

Object	Estimate

Lesson 2 Metric Units of Capacity 783

Problem Solving

11. Processes & Practices 2 **Reason** Jenna said that she took 4 milliliters of medicine when she was sick. Is this a reasonable statement? Explain.

_____

_____

12. Jonah said he drank 3 liters of water after his soccer game. Is this a reasonable statement? Explain.

_____

_____

_____

Brain Builders

13. Processes & Practices 4 **Model Math** List three objects in your house that can hold more than 1 liter. Estimate the capacity of each item.

_____

14. Processes & Practices 1 **Plan Your Solution** Suppose you have a 4-liter bucket and a 7-liter bucket. You need 5 liters of water for an aquarium. Explain how to get 5 liters of water if neither bucket is marked.

_____

_____

_____

_____

15. **Building on the Essential Question** How is measuring capacity with the metric system similar to measuring capacity with the customary system? How is it different?

_____

_____

Name

MY Homework

Lesson 2
Metric Units of Capacity

Homework Helper

Need help? connectED.mcgraw-hill.com

Thom is making stew for his family. Is it more reasonable to say the capacity of the stew pot is 5 liters or 5 milliliters?

You know that a milliliter is a tiny amount—about the capacity of an eyedropper.

You know that a liter is a greater amount—about the capacity of a large water bottle.

It would not be reasonable to estimate the capacity of a stew pot in milliliters.

So, it would be more reasonable to say the capacity of the stew pot is 5 liters.

Practice

Choose the most reasonable estimate for each capacity.

1.
   - Ⓐ 40 liters
   - Ⓑ 4 liters
   - Ⓒ 40 milliliters
   - Ⓓ 4 milliliters

2.
   - Ⓕ 10 mL
   - Ⓖ 100 mL
   - Ⓗ 10 L
   - Ⓘ 100 L

3.
   - Ⓐ 1 liter
   - Ⓑ 3 liters
   - Ⓒ 7 liters
   - Ⓓ 10 liters

4.
   - Ⓕ 17 mL
   - Ⓖ 170 mL
   - Ⓗ 170 L
   - Ⓘ 17 L

Lesson 2 My Homework 785

Match each object to its reasonable capacity.

5. bottle of nail polish • 300 liters

6. bathtub • 2 liters

7. large pitcher • 15 milliliters

Problem Solving

8. **Processes &Practices 1** **Check for Reasonableness** Emerson needs to use eye drops. Is it reasonable for her to put 1 milliliter of drops in each eye? Explain.

_____

9. Ryan fills his cat's water bowl. Is it reasonable to say he uses 1 milliliter of water? Explain.

_____

Brain Builders

10. Identify 2 objects you could find in a grocery store that hold less than 100 milliliters and 2 objects that hold more than 100 milliliters.

_____

_____

Vocabulary Check

Write a vocabulary term to complete each sentence.

liters   milliliters

11. The capacity of a baby's bottle would be measured in _____.

12. The capacity of a fish tank would be measured in _____.

13. **Test Practice** Which is a reasonable estimate for the capacity of a bottle of mouthwash?

Ⓐ 1 milliliter   Ⓒ 1 liter

Ⓑ 20 milliliters   Ⓓ 20 liters

Name _____

Lesson 3
Metric Units of Mass

ESSENTIAL QUESTION
How can conversion of measurements help me solve real-world problems?

Mass is the amount of matter an object has. The mass of an object is not affected by gravity. However, an object's weight differs depending on gravity.

gram (g)

The mass of a penny is about 1 gram.

kilogram (kg)

The mass of six medium apples is about 1 kilogram.

1,000 grams (g) = 1 kilogram (kg)

Math in My World

Example 1

Which is the more reasonable estimate for the mass of the laptop, 2 grams or 2 kilograms?

Use logic to estimate the mass.

2 grams

Helpful Hint
2 grams would have the same mass as about 2 pennies.

2 kilograms

Helpful Hint
2 kilograms would have the same mass as about 12 medium apples.

So, _____ is the more reasonable estimate.

Online Content at connectED.mcgraw-hill.com

Lesson 3   787

Example 2

Is it more reasonable to say that a rabbit's mass is 3 grams or 3 kilograms?

3 grams   3 kilograms

Helpful Hint
3 grams is too small.

Helpful Hint
3 kilograms is a reasonable estimate.

So, _____ is the more reasonable estimate.

Guided Practice

Circle the more reasonable estimate for each mass.

1.

   25 grams

   25 kilograms

2.

   450 grams

   450 kilograms

Talk MATH

Explain the difference between weight and mass.

788   Chapter 12 Metric Measurement

Name ....

Independent Practice

Circle the more reasonable estimate for each mass.

3.

8 grams

8 kilograms

4.

100 grams

100 kilograms

5.

25 grams

25 kilograms

6.

20 grams

20 kilograms

7.

30 grams

30 kilograms

8.

50 grams

50 kilograms

9. The table lists items that can be found in the classroom. Estimate the mass of each object. Record your estimates in the table.

Object	Estimate
glue bottle	
paper clip	
pencil	
stapler	

Lesson 3 Metric Units of Mass 789

Problem Solving

10. Tyler bought a large bag of peanuts at a baseball game. Is it more reasonable to say that the mass of the peanuts is 1 gram or 1 kilogram?

___

11. Processes & Practices 6 **Explain to a Friend** Alicia is buying 6 oranges that cost $1 per kilogram. Is it reasonable to say that the cost of the oranges will be greater than $6? Explain to a friend.

___

Brain Builders

12. Processes & Practices 4 **Model Math** List five classroom objects that have a mass greater than 1 kilogram. Estimate the mass of each.

___

13. Processes & Practices 2 **Reason** Items with greater mass cost more money to ship. Would it cost more to ship a box of 8 markers or a box of 20 paper clips? Explain.

___

14. Building on the Essential Question Describe a real-world situation where you would estimate the mass of something using a metric unit.

___

Name _____

MY Homework

Lesson 3

Metric Units of Mass

Homework Helper

Need help? connectED.mcgraw-hill.com

Geneva subscribes to a nature magazine. Is it reasonable to estimate that the mass of one issue of the magazine is 25 grams or 25 kilograms?

Twenty-five kilograms is too much.

You know that the mass of 1 penny is about 1 gram. Imagine holding 25 pennies in one hand and a magazine in the other. They would probably feel about the same.

So, it is reasonable to say the mass of one magazine is about 25 grams.

Practice

Circle the more reasonable estimate for each mass.

1.

1,500 grams   1,500 kilograms

2.

5 grams   5 kilograms

3.

3 grams   3 kilograms

4.

14 grams   14 kilograms

Complete the table by writing a reasonable unit of mass, grams or kilograms, for each object.

Mass of Fruits and Vegetables	
Object	Mass (g or kg)
5. grape	1 ☐
6. pumpkin	2 ☐
7. apple	150 ☐
8. cantaloupe	1 ☐
9. potato	1 ☐

Brain Builders

10. The mass of a pen cap is 1 unit. What metric unit, gram or kilogram, was used to measure the mass of the pen cap? Explain.

11. **Processes &Practices** **Explain to a Friend** Julio is buying a carton of blueberries that has a mass of 100 grams. Is it reasonable to say that there are 250 blueberries in the carton? Explain.

Vocabulary Check

Match each vocabulary term to its definition or example.

12. kilogram   • the amount of matter an object has

13. mass   • a metric unit of mass equal to about 1 penny

14. gram   • 1,000 grams

15. **Test Practice** Which is a reasonable estimate for the mass of a toothbrush?

Ⓐ 2 grams   Ⓑ 20 grams   Ⓒ 200 grams   Ⓓ 2,000 grams

Check My Progress

Vocabulary Check

1. Use the word bank to complete the charts about the **metric system** of measurement.

   centimeter gram kilogram kilometer mass

   milliliter millimeter meter liter

length

mass

capacity

Concept Check

2. Use the word bank to write each vocabulary word next to its abbreviation.

   mm _____   cm _____

   mL _____   km _____

   g _____   kg _____

   L _____   m _____

Check My Progress 793

Problem Solving

3. Adrianna went on a hiking trip. Which measurement best describes how far she hiked, 10 kilometers or 10 meters?

   _____

4. Which is the more reasonable estimate for the mass of a dog, 20 grams or 20 kilograms?

   _____

Brain Builders

5. Raul has a bottle of salad dressing. Is 700 milliliters or 700 liters a more reasonable estimate for the capacity of the bottle of salad dressing? Give an example of something that represents the other capacity.

   _____

   _____

6. **Test Practice** Which of the following holds about 800 milliliters of water?

   Ⓐ

   Ⓑ

   Ⓒ

   Ⓓ

Name ....

Lesson 4
Problem-Solving Investigation
STRATEGY: Make an Organized List

ESSENTIAL QUESTION
How can conversion of measurements help me solve real-world problems?

Learn the Strategy

Sebastian has 0.24 of a dollar. How many different combinations of coins could he have?

1 Understand

What facts do you know?

Sebastian has _____ of a dollar.

What do you need to find?

the number of possible coin combinations

2 Plan

I will make an organized list to solve the problem.

3 Solve

0.24 of a dollar = 24¢

- 2 dimes, 4 pennies
- 1 dime, 2 nickels, 4 pennies
- 1 dime, 1 nickel, 9 pennies
- 4 nickels, 4 pennies
- 3 nickels, 9 pennies
- 2 nickels, 14 pennies
- 24 pennies
- 1 dime, 14 pennies
- 1 nickel, 19 pennies

There are _____ different combinations.

4 Check

Does your answer make sense? Explain.

Online Content at connectED.mcgraw-hill.com

Lesson 4 795

Practice the Strategy

Brody has three cats. One has a mass of 4,523 grams. One has a mass of 5,012 grams. One has a mass of 4,702 grams. If Brody picks up two of the cats at once, what are the possible total masses that Brody is carrying?

 Understand

What facts do you know?
_____
_____

What do you need to find?
_____
_____

 Plan

_____
_____

 Solve

 Check

Does your answer make sense? Explain.
_____
_____

796 Chapter 12 Metric Measurement

Apply the Strategy

Solve each problem by making an organized list.

1. **Processes & Practices 1** **Make a Plan** Brianna has 0.16 of a dollar. How many different combinations of coins could she have?

2. There were three races at the track meet. The distances were 100 meters long, 800 meters long, and 3,200 meters long. Suppose Lucy ran two of the races. What are the possible total distances that she ran?

Brain Builders

3. Aaron has 3,700 milliliters of lemonade in a pitcher. He has three cups. Their capacities are 320 milliliters, 495 milliliters, and 583 milliliters. Suppose Aaron fills two of the cups. What are the possible capacities of lemonade that he could have left in the pitcher?

4. A museum elevator has a weight limit of 1,200 kilograms. Ellen wants to take the following sculptures on the elevator: 510 kilograms, 725 kilograms, 830 kilograms, 600 kilograms. What is the minimum number of trips Ellen needs to take? Explain.

5. Dean has four pieces of clay to make a clay pot. The masses of the pieces are 10 grams, 15 grams, 20 grams, and 14 grams. If he uses two pieces of clay for each pot, how could he make two pots that have close to the same mass?

Lesson 4 Problem-Solving Investigation 797

Review the Strategies

Use any strategy to solve each problem.
- Make an organized list.
- Guess, check, and revise.
- Find extra or missing information.
- Use logical reasoning.

6. There are three trees in a backyard. The second tree is half as tall as the first. The third tree is taller than the second tree and shorter than the first tree. The total height of the trees is 24 feet. What is the height of each tree?

7. There are three lines. The first line is 3 times as long as the second. The second line is 4 meters longer than the third. The third line is 2 meters long. How long is the first line?

8. Darin has 5 coins that total 62¢. What are the coins?

9. Processes & Practices 4 Model Math Alfonso, Erik, Owen, and Alek are going hiking in pairs. How many different pairs of hiking partners are possible? List them.

798 Chapter 12 Metric Measurement

Name ....

MY Homework

Lesson 4

Problem Solving: Make an Organized List

Homework Helper

Need help? connectED.mcgraw-hill.com

Quinn's backpack can carry **5 kilograms of mass.** Look at the items in the chart. What possible combinations of items can Quinn carry in her backpack without exceeding 5 kilograms?

Item	Mass
math book	3 kg
art supply kit	2 kg
lunch box	2 kg
water bottle	1 kg

1 Understand

What facts do you know?

Quinn's backpack can carry 5 kilograms of mass.

What do you need to find?

I need to find the possible combinations of items Quinn can carry in her backpack.

2 Plan

I will make an organized list of the possible combinations

3 Solve

- math book, art supply kit—5 kg
- math book, lunch box—5 kg
- math book, water bottle—4 kg
- art supply kit, lunch box, water bottle—5 kg
- art supply kit, lunch box—4 kg
- art supply kit, water bottle—3 kg
- lunch box, water bottle—3 kg

So, there are seven possible combinations.

4 Check

Does the answer make sense?

Yes. I listed each combination and its total mass. None of the masses are greater than 5 kilograms.

So, the answer is reasonable.

Lesson 4 My Homework 799

Problem Solving

Solve each problem by making an organized list.

1. Paul's bathtub is clogged. He has to empty 30 liters of water by hand. Paul has a 3-liter, a 4-liter, and a 5-liter bucket. If Paul carries two buckets each trip, what combinations of sizes allow him to empty the bathtub in exactly four trips?

   _____

   _____

2. Tyra is training for a bicycle race. Each week she rides a total distance greater than 10 kilometers and less than or equal to 30 kilometers. If the distance is always an even number and a multiple of 3, what are the possible distances Tyra rides in one week?

   _____

Brain Builders

3. **Processes & Practices 1** **Keep Trying** Lexi's bulletin board is 40 centimeters wide. Each of her ribbons is 4 centimeters wide, and her photos are 12 centimeters wide. What combinations of ribbons and photos will fit side by side with no overlap on Lexi's bulletin board? Draw a diagram to support your answers.

   _____

   _____

4. Carmen buys a pack of crackers for 75 cents from the vending machine. She puts a $1-bill in the machine. What combination of coins, excluding pennies, could Carmen get in change? If pennies were included, would there be more or less than an additional 10 combinations?

   _____

   _____

Name ....

Lesson 5
Convert Metric Units

ESSENTIAL QUESTION

How can conversion of measurements help me solve real-world problems?

You can multiply to convert, or change between, units.

 Math in My World

Example 1

The tree in Camryn's front yard is 4 meters tall. How many centimeters tall is the tree?

Since meters are larger than centimeters, multiply.

4 × 100 = 400

Multiply by 100 because there are 100 centimeters in each meter.

4 meters = _____ centimeters

So, the tree is _____ centimeters tall.

Example 2

Complete. 5 liters = ■ milliliters

Since liters are larger than milliliters, multiply.

5 × 1,000 = 5,000

Multiply by 1,000 because there are 1,000 milliliters in each liter.

So, 5 liters = _____ milliliters.

Metric Units of Length

1 centimeter (cm) = 10 millimeters (mm)
1 meter (m) = 100 centimeters (cm)
1 kilometer (km) = 1,000 meters (m)

Metric Units of Capacity

1 liter (L) = 1,000 milliliters (mL)

Online Content at connectED.mcgraw-hill.com

Example 3

Convert 7 kilograms to grams.

7 kilograms = ? grams

Metric Units of Mass
1 kilogram (kg) = 1,000 grams (g)

Kilograms are larger than grams. So, use multiplication.

Multiply by 1,000 because 1 kilogram = 1,000 grams.

1,000 × 7 = _____

So, 7 kilograms = _____ grams.

Guided Practice

Complete each conversion table.

List the number pairs in the last column of the conversion table.

1.
kilometers (km)	meters (m)	(km, m)
1	1,000	(1, 1,000)
2		
3		
4		

2.
centimeters (cm)	millimeters (mm)	(cm, mm)
1		
2		
3		
4		

3.
meters (m)	centimeters (cm)	(m, cm)
5		
6		
7		
8		

Explain why multiplication is used to convert from a larger unit to a smaller unit.

4.
liters (L)	milliliters (mL)	(L, mL)
1	1,000	(1, 1,000)
2		
3		
4		

Name ....

Independent Practice

Complete each conversion table.

5.

meters (m)	centimeters (cm)	(m, cm)
4		
5		
8		
9		

6.

kilograms (kg)	grams (g)	(kg, g)
7		
9		
11		
13		

Algebra Find each unknown number.

7. 6 L = ■ mL

   ■ = _____

8. 5 m = ■ cm

   ■ = _____

9. 2 kg = ■ g

   ■ = _____

10. 5 cm = ■ mm

    ■ = _____

11. 12 kg = ■ g

    ■ = _____

12. 4 m = ■ mm

    ■ = _____

13. 5 L = ■ mL

    ■ = _____

14. 7 km = ■ m

    ■ = _____

15. 19 m = ■ cm

    ■ = _____

16. 9 kg = ■ g

    ■ = _____

17. 18 L = ■ mL

    ■ = _____

18. 22 cm = ■ mm

    ■ = _____

19. How many times larger is one kilogram than one gram? _____

20.   **Use Number Sense** How many times longer is one kilometer than one meter? _____

21. How many times longer is one meter than one centimeter? _____

Lesson 5 Convert Metric Units **803**

Problem Solving

22. The mass of Kendall's bicycle is 12 kilograms. What is the mass of the bicycle in grams?

23. Mrs. Liu's house is 7 meters tall. How tall is the house in centimeters?

24. Processes & Practices 2 **Use Number Sense** Javier needs 2 liters of iced tea for a picnic. How many milliliters of iced tea does he need?

Brain Builders

25. Avery's dad is running a race that is 6 kilometers long. How many centimeters is that race?

26. Processes & Practices 3 **Which One Doesn't Belong?** Circle the measurement that does not belong. Explain.

| 300 grams | 10 kilograms | 10 pounds | 600 grams |

27. Building on the Essential Question When converting from a larger unit to a smaller unit, what happens to the number of units in the measurement? Explain.

804 Chapter 12 Metric Measurement

Name _____

MY Homework

Lesson 5
Convert Metric Units

Homework Helper

Need help? connectED.mcgraw-hill.com

Liam is delivering trophies for the soccer team. The mass of one trophy is 2 kilograms. What is the mass in grams of the trophy?

Kilograms are larger than grams, so you will multiply.

2 kilograms × 1,000 = 2,000 grams

So, the mass of the trophy in grams is 2,000 grams.

Helpful Hint
1 centimeter (cm) = 10 millimeters (mm)
1 meter (m) = 100 centimeters (cm)
1 kilometer (km) = 1,000 meters (m)
1 liter (L) = 1,000 milliliters (mL)
1 kilogram (kg) = 1,000 grams (g)

Kymie said her driveway is 14 meters long. How long is her driveway in centimeters?

Meters are larger than centimeters, so you will multiply.

14 meters × 100 = 1,400 centimeters

So, Kymie's driveway is 1,400 centimeters long.

Practice

Algebra Find each unknown number.

1. 7 kg = ■ g

■ = _____

2. ■ mm = 9 cm

■ = _____

3. 5 L = ■ mL

■ = _____

4. 23 m = ■ cm

■ = _____

5. 17 kg = ■ g

■ = _____

6. 450 cm = ■ mm

■ = _____

Problem Solving

7. Molly measured the distance her paper airplane flew. The paper airplane traveled 5 meters. How many centimeters did the paper airplane travel?

8. Thad is riding his bike to the library, which is 3 kilometers away. How many meters away is the library?

9. Patrick has 5 liters of water. How many milliliters of water does he have?

Brain Builders

10. Maria's suitcase has a mass of 14 kilograms. Her sister's suitcase has a mass of 1,600 grams. Which suitcase has a greater mass? Explain.

11. **Processes & Practices 2** **Use Number Sense** Minh is packing books into boxes. The mass of one of the boxes is 20 kilograms. If the box holds 10 books that each have the same mass, what is the mass of each book in grams?

12. **Test Practice** Which is equivalent to 300 meters?

    Ⓐ 30 kilometers
    Ⓑ 3 kilometers
    Ⓒ 30,000 centimeters
    Ⓓ 3,000 centimeters

Name _____

Lesson 6
Solve Measurement Problems

ESSENTIAL QUESTION
How can conversion of measurements help me solve real-world problems?

 Math in My World

Example 1

Lauren lives 0.2 kilometer from Alex. Alex lives three times as far from Colin's house than Lauren's house. How far away does Alex live from Colin?

Find 3 × 0.2.

You can use a number line to solve the problem.

Start at zero. Count by 0.2 three times.

3 × 0.2 = 0.6

So, Alex lives _____ kilometer from Colin.

Check

Convert 0.2 to a fraction. Then multiply the fraction by 3.

$0.2 = \text{two tenths} = \frac{2}{10}$

$3 \times \frac{2}{10} = 3 \times \left(2 \times \frac{1}{10}\right)$ $\frac{2}{10} = 2 \times \frac{1}{10}$

$\phantom{3 \times \frac{2}{10}} = (3 \times 2) \times \frac{1}{10}$ Associative Property

$\phantom{3 \times \frac{2}{10}} = 6 \times \frac{1}{10}$ Multiply. 3 × 2 = 6

$\phantom{3 \times \frac{2}{10}} = \frac{6}{10}$ 6 groups of $\frac{1}{10}$ is $\frac{6}{10}$.

Helpful Hint
Think of $\frac{2}{10}$ as a multiple of $\frac{1}{10}$.

Since $\frac{6}{10} = \text{six tenths} = 0.6$, the answer is correct.

Online Content at connectED.mcgraw-hill.com

Example 2

Javier poured 500 milliliters of lemon juice and 2 liters of water in a pitcher to make lemonade. How many milliliters of lemon juice and water did he pour into the pitcher in all?

 Convert.

Convert 2 liters to milliliters.

Since 1 liter = 1,000 milliliters, multiply the number of liters by 1,000.

2 × 1,000 = 2,000

So, 2 liters = 2,000 milliliters.

 Add.

2,000 milliliters + 500 milliliters = 2,500 milliliters

So, Javier poured _____ milliliters of lemon juice and water into the pitcher.

Guided Practice

1. Evelyn is in a relay race with three other runners. Each runner runs 0.1 kilometer. What is the total distance run by all four runners? Use the number line.

   _____

   Talk MATH
   Explain how you can check your answer for Exercise 1.

2. A bag of potatoes has a mass of 4 kilograms. Some potatoes are taken out. The mass is now 2,305 grams. What is the mass of the potatoes in grams that were taken out of the bag?

   _____

808 Chapter 12 Metric Measurement

Name _____

Independent Practice

3. A ribbon is 1 meter long. Keira cuts off a piece of the ribbon that is 0.4 meter long. How much of the ribbon is left? Use the number line.

_____

4. There are two books. Each has a mass of 0.4 kilogram. What is the total mass of the two books? Use the number line.

The total mass is _____.

Convert to solve each problem. Draw a number line if needed.

5. One insect is 47 millimeters long. Another insect is 3 centimeters long. What is the total length in millimeters of the insects?

6. A table has a mass of 7 kilograms. A chair has a mass of 4,048 grams. What is the total mass in grams of the table and the chair?

Lesson 6 Solve Measurement Problems 809

Problem Solving

7. Cole has a plastic cup that has 125 milliliters of water in it. He drinks 37 milliliters of the water. How much water is left in the cup?

_____

8. **Processes &Practices 3** **Justify Conclusions** Sam had 0.3 of a dollar. Then he found four nickels and a penny. Does he have enough money to buy something that costs 50¢? Explain.

_____

_____

Brain Builders

9. Each sports bag can hold 6 kilograms of equipment. The golf balls have a total mass of 3,402 grams. The hockey pucks have a total mass of 2,932 grams. How many sports bags will be needed to hold the equipment? Explain.

_____

_____

10. **Processes &Practices 2** **Use Symbols** Compare. Write <, >, or =.

    3 L + 2,492 mL ◯ 2 L + 1,301 mL + 2,191 mL

11. **Building on the Essential Question** How do I know when it is necessary to convert units before solving a problem? Give an example.

_____

_____

_____

Name ....

Lesson 6

Solve Measurement Problems

Homework Helper   Need help? connectED.mcgraw-hill.com

Alfredo has a board that is 2 meters long. He needs 3 pieces that are each 50 centimeters long. How many centimeters of the original board will be left after he cuts the 3 pieces he needs?

1 Convert the meters to centimeters. Meters are larger than centimeters, so you will multiply.
2 meters × 100 = 200 centimeters

2 Multiply to find the total length of the 3 pieces Alfredo needs.
3 × 50 centimeters = 150 centimeters

3 Subtract to find the new length of the original board.
200 − 150 = 50 centimeters

So, Alfredo will have 50 centimeters left of the original board.

Practice

1. Chloe buys 3 jars of peanut butter for the food drive. Each jar has a mass of 0.2 kilogram. What is the mass of all 3 jars of peanut butter? Use the number line.

Lesson 6 My Homework 811

Problem Solving

2. Aidan mixes 630 milliliters of blue paint with 2 liters of red paint to make purple paint. How many milliliters of purple paint does Aidan have?

3. Celine is knitting a scarf. The finished length will be 1.2 meters. So far she has knitted 0.8 meters. How many more meters does Celine need to knit? Draw a number line to solve.

4. Joanie's bag of trail mix has a mass of 0.75 kilogram. She and some friends eat 0.5 kilogram of the mix. How many kilograms of trail mix are left? Draw a number line to solve.

Brain Builders

5. **Processes & Practices 5** **Use Math Tools** Ross found 5 ladybugs and 1 caterpillar. Each ladybug is 8 millimeters long and the caterpillar is 3 centimeters long. If he lays the ladybugs and caterpillar in a row, what is the total length in millimeters? Draw a number line to solve.

6. Lucas is riding his bike to the park, which is 2 kilometers from his house. When he is one-fourth of the way to the park, it starts to rain. Lucas turns around and rides back home. How many meters did Lucas ride? Explain.

7. **Test Practice** Grayson stacked 17 kilograms of firewood on Saturday and 15,000 grams of firewood on Sunday. How many more grams of firewood did he stack on Saturday?

   Ⓐ 32,000 grams   Ⓒ 14,983 grams

   Ⓑ 15,017 grams   Ⓓ 2,000 grams

Review

Chapter 12
Metric Measurement

Vocabulary Check

Draw a line to the sentence that each vocabulary word completes.

1. capacity
2. convert
3. mass
4. metric system
5. length

- Centimeter, gram, and liter are all examples of units of measure from the _____.
- _____ is the amount of liquid a container holds.
- _____ is the measurement of a line between two points.
- When you change the unit of measure, you _____ measurements.
- The amount of matter that an object has is known as its _____.

Color the metric units of length red. Color the metric units of capacity blue. Color the metric units of mass green. Then write each abbreviation on the roof.

6. liter

7. millimeter

8. kilogram

9. centimeter

10. milliliter

11. gram

12. meter

13. kilometer

Concept Check

14. Measure the width of the flower to the nearest centimeter.

Length: _____

15. Choose the best estimate for the length of a bottle of glue.

Ⓐ 15 millimeters

Ⓑ 15 centimeters

Ⓒ 15 meters

Ⓓ 15 kilometers

16. Circle the more reasonable estimate for the capacity of a bucket.

6 milliliters

6 liters

17. Circle the more reasonable estimate for the mass of a chair.

15 grams

15 kilograms

Complete the conversion table.

18.

kilograms (kg)	grams (g)	(kg, g)
12		
14		
16		
18		

19. How many times longer is one kilometer than one meter? _____

20. How many times longer is one centimeter than one millimeter? _____

Problem Solving

21. Carson has 0.21 of a dollar. How many different combinations of coins could he have?

22. There are three picture frames. Each has a mass of 0.2 kilogram. What is the total mass of the three picture frames? Use the number line.

Brain Builders

23. Justin drank 1 liter of water during soccer practice. He drank 2,000 milliliters of water during the soccer game. How many milliliters of water did he drink during the practice and the game? Explain.

24. Julian walked a distance of 2 meters. Keira walked a distance of 300 centimeters. Which student walked farther? How much farther?

25. Test Practice Henry's water bottle has a capacity of 1 liter. How many milliliters will Henry drink if he drinks 1 bottle each day for 10 days?

Ⓐ 1 milliliter

Ⓒ 1,000 milliliters

Ⓑ 10 milliliters

Ⓓ 10,000 milliliters

My Chapter Review 815

Reflect

Chapter 12

Answering the **ESSENTIAL QUESTION**

Use what you learned about metric measurements to complete the graphic organizer.

Length
Real-World Example

ESSENTIAL QUESTION
How can conversion of measurements help me solve real-world problems?

Capacity
Real-World Example

Mass
Real-World Example

Reflect on the **ESSENTIAL QUESTION** Write your answer below.

_____
_____
_____

Name _____   Date _____

Score _____

Performance Task

Let's Take the Ferry

A ferry operates between two parts of town, going across the lake. The ferry carries people and cars.

Show all your work to receive full credit.

Part A

Alan's car was the first one on the ferry for the one-hour trip. Choose the best measure for each item in his car.

Item	Measures			
The capacity of a cup in the cupholder	3 milliliters	300 milliliters	3 liters	30 liters
The mass of the spare tire	25 grams	500 grams	12 kilograms	50 kilograms
The length of the car	30 centimeters	20 meters	3 meters	8 kilometers

Part B

The Coffee Corner on the ferry makes coffee in an urn that holds 6 liters. How many of these containers can be filled from the urn **in order** before they will need to make more coffee? Explain.

250 mL	250 mL	350 mL	250 mL	250 mL
250 mL	250 mL	350 mL	800 mL	350 mL
250 mL	250 mL	350 mL	250 mL	250 mL
350 mL	250 mL	250 mL	800 mL	350 mL
250 mL	350 mL	250 mL	250 mL	350 mL
250 mL	250 mL	350 mL	250 mL	250 mL

Online Content at connectED.mcgraw-hill.com

Part C

On the ferry, each row of cars and trucks is 16 meters. The table shows the lengths of the cars and trucks that are lined up to board the ferry. Which row, or rows, of cars and trucks will fit? Explain. Show your work.

Row 1:

3.5 meters | 3.1 meters | 3.2 meters | 4 meters | 2.8 meters

Row 2:

3.2 meters | 300 centimeters | 6 meters | 3.3 meters

Row 3:

3.5 meters | 2.4 meters | 3.8 meters | 280 centimeters | 3.3 meters

Row 4:

630 centimeters | 3.8 meters | 4 meters | 2.9 meters

Chapter 13 Perimeter and Area

ESSENTIAL QUESTION

Why is it important to measure perimeter and area?

Let's Cheer OUR TEAM!

Watch a video!

817

Name _____

MY Chapter Project

It's Your Area

1. Write the name of a building in your community you would like to learn about:
_____

2. Find what shapes were used to create the floor plan of the building. Draw a sketch of the floor plan below:

3. Using the Internet or other resources, find the actual length and width of each section of the floor. Use the measurements to calculate the area of each section.

Section of Building	Length (feet)	Width (feet)	Area (square feet)

4. Add the areas of all the sections to find the total area of the building.

The total area of the building is _____ square feet.

Name ....

Draw an array to model each multiplication problem. Find the product.

1. 4 × 8 = _____

2. 7 × 7 = _____

3. 3 × 12 = _____

4. 10 × 4 = _____

5. 9 × 11 = _____

6. 4 × 15 = _____

7. Cora is making a picture. It will be a square. How many sides are the same length?

_____

Shade the boxes to show the problems you answered correctly.

How Did I Do? 1 2 3 4 5 6 7

Online Content at connectED.mcgraw-hill.com

MY Math Words

Review Vocabulary

length   product

Making Connections

Use the review vocabulary to complete the chart.

Look at the array below. Write a multiplication problem to describe it. Then find the product.	What number represents the length of this array?
Draw an array that represents a product of 20.	Suppose the array below represents a figure. Describe this figure's sides.

820 Chapter 13 Perimeter and Area

MY Vocabulary Cards

Lesson 13–3

area

32 square feet

Lesson 13–1

perimeter

$P = \ell + w + \ell + w$
$P = (2 \times \ell) + (2 \times w)$

Lesson 13–3

square unit

Length (ft)	Width (ft)	Area (sq ft)
7	3	21

Lesson 13–3

unit square

1 unit

Ideas for Use

- Write a tally mark on each card every time you read or write the word. Challenge yourself to use at least 3 tally marks for each card.
- Use the blank cards to review problem-solving strategies, such as work backward, make a table, or draw a picture.

The distance around a closed figure.

Would it be possible to double the lengths and sides of any figure to find its perimeter? Explain.

The number of square units needed to cover a figure without any overlap.

Draw a rectangle. Label the sides' lengths. Exchange papers with a friend to find the area of the rectangle.

A square with a side length of one unit.

A friend tells you he used grid paper to draw a rectangle with a length of 12 unit squares and a width of 5 unit squares. What is the rectangle's total number of unit squares?

A unit for measuring area. Contains one unit square.

Explain how using square units to measure is different from using units.

MY Foldable

 Follow the steps on the back to make your Foldable.

Perimeter	Draw	Area
$2 + 5 + 2 + 5 = 14$ $P = 14$ ft	2 ft, 5 ft rectangle	$2 \times 5 = 10$ $A = 10$ sq ft
$20 + 23 + 20 + 23 = 86$ $P = 86$ in.		
	3 cm, 11 cm rectangle	$26 \times 4 = 104$ $A = 104$ sq in.

Perimeter | Draw | Area

$21 + 21 + 21 + 21 = 84$

$P = 84$ ft

21 ft
21 ft

$21 \times 21 = 441$

$A = 441$ sq ft

$60 \times 60 = 3{,}600$

$A = 3{,}600$ sq yd

1 cm
1 cm

$34 + 34 + 34 + 34 = 136$

$P = 136$ in.

Name _____

Lesson 1
Measure Perimeter

ESSENTIAL QUESTION
Why is it important to measure perimeter and area?

The distance around a closed figure is called the **perimeter**.

Math in My World

Example 1

Berto walked around a park on the rectangular path shown. How far did Berto walk?

6 yd

12 yd

Opposite sides of a rectangle are equal. So, the side lengths are 12 yards, 12 yards, 6 yards, and 6 yards.

Add the measures of all of the sides of the figure.

Perimeter = 12 yards + 12 yards + 6 yards + 6 yards

Perimeter = _____ yards

So, Berto walked _____ yards.

Key Concept — Perimeter of a Rectangle

Words   To find the perimeter of a rectangle, add the lengths of the sides. The perimeter of a rectangle also equals 2 times its length plus 2 times its width.

Symbols $P = \ell + w + \ell + w$
$P = (2 \times \ell) + (2 \times w)$

Online Content at connectED.mcgraw-hill.com

A square has four sides of equal length. To find the perimeter of a square, multiply the length of one side by four.

Example 2

Find the perimeter of a square with side lengths of 6 inches.

One Way Add the measures of all of the sides of the figure.

$P = 6$ inches $+ 6$ inches $+ 6$ inches $+ 6$ inches

$P = 24$ inches

Another Way Use a formula.

Multiply the length of one side by 4 because there are 4 sides of equal length.

$P = 4 \times s$ ← s is equal to one side length

$P = 4 \times 6$ inches

$P = 24$ inches

So, the perimeter of the square is _____ inches.

Guided Practice

Find each perimeter.

1.

   $P =$ _____

2.

   $P =$ _____

Talk MATH

Describe two ways to find the perimeter of a square.

826 Chapter 13 Perimeter and Area

Name

Independent Practice

Find each perimeter.

3.

   P = _____

4.

   P = _____

5.

   P = _____

6.

   P = _____

Algebra Find the unknown side length. Write an equation to find the perimeter.

7.

   The unknown side length is _____.

   _____

8.

   The unknown side length is _____.

   _____

Find the perimeter of each rectangle in units.

9.

   P = _____

10.

    P = _____

Lesson 1 Measure Perimeter 827

Problem Solving

Use the picture of a football field for Exercises 11–12.

11. What is the perimeter of the football field without the end zones?

_____

12. Each end zone is 10 yards long. What is the perimeter of the entire football field in feet?

_____

Brain Builders

13. **Processes &Practices 2** **Use Number Sense** A baseball diamond is shaped like a square. The perimeter is 360 feet. What is the length of each side? Explain.

_____

_____

14. **Processes &Practices 6** **Explain to a Friend** Explain to a friend how to find the perimeter of the figure shown to the right. Then find the perimeter.

_____

_____

_____

15. **Building on the Essential Question** How can a formula help me to find perimeter?

_____

_____

_____

828 Chapter 13 Perimeter and Area

Name _____

MY Homework

Lesson 1

Measure Perimeter

Homework Helper

Need help? connectED.mcgraw-hill.com

Claire plans to glue ribbon around the edges of the picture frame. How much ribbon will she need?

One Way Add the measures of all sides of the figure.

You know that opposite sides of a rectangle are equal, so the measures are 5 inches, 5 inches, 8 inches, and 8 inches.

P = 5 inches + 5 inches + 8 inches + 8 inches
P = 26 inches

Another Way Use a formula.

P = (2 × 8 inches) + (2 × 5 inches)
P = 16 inches + 10 inches
P = 26 inches

So, Claire will need 26 inches of ribbon.

Helpful Hint
The perimeter of a rectangle equals 2 times the length plus 2 times the width.

$P = (2 \times \ell) + (2 \times w)$

Practice

Find each perimeter.

1. P = _____ cm

2. P = _____ yd

3. P = _____ mm

4. P = _____ ft

Lesson 1 My Homework 829

Find the perimeter of each rectangle in units.

5.

P = _____ units

6.

P = _____ units

Brain Builders

7. **Processes &Practices 4** **Model Math** Justin walked his dog 2 blocks west, 6 blocks north, 2 blocks east, and 6 blocks south. Then he walked his other dog along the same route. Draw a picture to represent Justin's route. Write and simplify an expression to show the number of blocks Justin walked in all.

_____

_____

8. A rectangle has a perimeter of 30 inches. One side measures 5 inches. What are the measures of the other three sides? Explain.

_____

_____

_____

9. Carolla is putting a border around the edge of a rectangular bulletin board. One side of the bulletin board measures 2 feet and another side measures 4 feet. Is 10 feet of border enough? Explain.

_____

_____

Vocabulary Check

10. Write a definition for the term perimeter. _____

_____

11. **Test Practice** What is the measure of the unknown side of the rectangle?

Ⓐ 23 meters   Ⓒ 7 meters

Ⓑ 14 meters   Ⓓ 5 meters

9 m, P = 32 m, ? m

Name _____

Lesson 2
Problem-Solving Investigation
STRATEGY: Solve a Simpler Problem

ESSENTIAL QUESTION
Why is it important to measure perimeter and area?

Learn the Strategy

It takes Jasmine 2 minutes to ride her bike one block. How long does it take Jasmine to ride the route shown three times?

1 Understand

What facts do you know?

It takes Jasmine _____ minutes to ride her bike 1 block.

What do you need to find?

How long it will take to ride the route three times

2 Plan

I will solve a simpler problem to find the answer.

3 Solve

$2 + 2 + 2 + 2 + 4 + 4 = 16$ ← total number of blocks Jasmine rides one time around

Jasmine rides 16 + 16 + 16, or 48, blocks when she rides three times around.

$\underset{\text{minutes per block}}{2} \times \underset{\text{total blocks}}{48} = \underset{\text{total minutes}}{96}$ ← number of minutes it takes her to ride three times around

So, it takes Jasmine _____ minutes to ride three times around.

4 Check

Does your answer make sense? Explain.

_____

_____

Online Content at connectED.mcgraw-hill.com

Lesson 2 831

Practice the Strategy

Hayden and his friends want to draw a court on the playground to play Four Square. The sides of each of the small squares are 4 feet. What is the perimeter of the entire court?

1 Understand

What facts do you know?

_____

What do you need to find?

_____

2 Plan

_____

3 Solve

4 Check

Does your answer make sense? Explain.

_____

Name

Apply the Strategy

Solve each problem by solving a simpler problem.

1. Clarissa has four pictures that are each the size of the one shown. What will be the perimeter of the rectangle formed if the four pictures are laid end to end as shown?

   5 in.

   3 in.

2. Mr. and Mrs. Lopez are putting square tiles on the floor in their bathroom. They can fit 6 rows of 4 tiles in the bathroom. How many tiles do they need to buy? If each tile costs $5, what is the total cost?

Brain Builders

3. Ling is putting up a wallpaper border on three walls that are each 14 feet long and 12 feet tall. The cost of the wallpaper border is $2 per foot. How much will it cost Ling to put wallpaper border along the top of the walls?

4. **Processes & Practices 1** **Plan Your Solution** Takeisha is placing placemats around a table. Each placemat is 14 inches long and 10 inches wide. The rectangular table is 42 inches long and 48 inches wide. What is the greatest number of placemats that Takeisha can place around the table without overlapping?

Lesson 2 Problem-Solving Investigation 833

Review the Strategies

Use any strategy to solve each problem.
- Solve a simpler problem.
- Draw a picture.
- Make a table.
- Guess, check, and revise.

5. **Processes & Practices 8** **Look for a Pattern** What is the perimeter of the eighth figure if this pattern continues?

6. Marcos is making three tile pictures. He uses 310 green tiles to make each picture. He uses 50 fewer red tiles than green tiles for each picture. How many red and green tiles does he use in all?

7. A type of bacteria doubles in number every 12 hours. After 2 days, there are 48 bacteria. How many bacteria were there at the beginning of the first day?

8. Martell is stacking boxes. Each box is 2 feet high. If he stacks 3 boxes on top of a table that is 3 feet high, what will be the total height of the boxes and the table?

9. Mr. Waters is building a rectangular sandbox. He needs to buy enough wood to go around the perimeter of the sandbox. If the length is 6 feet and the width is 5 feet, how many feet of wood does Mr. Waters need to buy?

834 Chapter 13 Perimeter and Area

Name

MY Homework

Lesson 2

Problem Solving:
Solve a Simpler Problem

Homework Helper

Need help? connectED.mcgraw-hill.com

Aaliyah and her 5 friends are picking up litter at the park. Each will clean a rectangular area with one side that measures 2 yards and another side that measures 6 yards. What is the total perimeter of all 6 areas they will clean?

1 Understand

What facts do you know?

Six people will each clean an area that measures 2 yards × 6 yards.

What do you need to find?

the total perimeter of all 6 areas

2 Plan

Solve a simpler problem.

3 Solve

Make a drawing and find the perimeter for one area. One side measures 2 yards and another side measures 6 yards.

Perimeter = 2 yd + 2 yd + 6 yd + 6 yd = 16 yd

Multiply 16 by the number of areas.

6 areas × 16 yards = 96 yards

So, the total perimeter of all six areas is 96 yards.

4 Check

Use addition to check the answer.

16 yd + 16 yd + 16 yd + 16 yd + 16 yd + 16 yd = 96 yards

So, the answer is correct.

Problem Solving

Solve each problem by solving a simpler problem.

1. There are two identical figures with all sides of equal length. The combined perimeter of the figures is 80 centimeters. What shape are the figures? What is the length of one side?

2. Miranda packed 19 glass ornaments into each box. She filled 5 boxes. What is the total number of ornaments in the boxes?

Brain Builders

3. Carrie is making a bedskirt for her twin beds. One side of a mattress measures 39 inches, and another side measures 75 inches. How many feet of fabric will Carrie need to make skirts for both beds? Explain.

4. Dalton's dad rode his bike 1 kilometer north, 2 kilometers west, 1 kilometer south, and 2 kilometers east. Dalton rode the same path 5 times. How many kilometers farther did Dalton ride?

5. **Processes & Practices 1 Make Sense of Problems** A square chessboard is made up of 64 white and black squares arranged in 8 rows and 8 columns. The side length of each square is 2 inches. What is the perimeter of the chessboard?

Check My Progress

Vocabulary Check

The distance around a closed figure is called the **perimeter**.

1. Which of the following is a formula for finding the perimeter of a rectangle? Circle the correct response.

$$P = \ell + w$$

$$P = 4 \times \ell \times w$$

$$P = \ell \times w$$

$$P = (2 \times \ell) + (2 \times w)$$

Concept Check

Find each perimeter.

2.

P = _____

3.

P = _____

4.

P = _____

5.

P = _____

Check My Progress 837

Problem Solving

6. Byron made a drawing of his room. His drawing is shown. What is the perimeter of Byron's room?

7. What is the perimeter of a square with side lengths of 4 inches?

Brain Builders

8. Felicia is building a rectangular garden. The garden will have a perimeter of 20 meters. Give three pairs of possible side lengths.

   Is it possible for her garden to be a square? Explain.

9. A rectangular poster has a length of 24 inches, and it is twice as long as it is wide. What is the perimeter of the poster?

10. **Test Practice** Which of the following is the perimeter of a square with whole-number side lengths?

    Ⓐ 10 centimeters   Ⓒ 20 centimeters

    Ⓑ 15 centimeters   Ⓓ 25 centimeters

Name _____

Lesson 3
Hands On
Model Area

ESSENTIAL QUESTION
Why is it important to measure perimeter and area?

A square with a side length of one unit is called a **unit square**.

A unit square has one **square unit** of area, and can be used to measure area. **Area** is the number of square units needed to cover a figure without overlapping.

Shading, or covering, a unit square results in one square unit.

Draw It

Find the area of the rectangles shown in the table.

Rectangle	Length (units)	Width (units)	Area (sq units)
▭▭▭▭			
(2×4 grid)			
(3×3 grid)			
(3×5 grid)			

 Draw each rectangle.
Use grid paper to draw each rectangle.

 Find the length and width of each rectangle.
Count the number of unit squares that cover the length and width of the rectangle. Record each in the table.

 Determine the area of each rectangle.
Count the number of whole squares that cover the rectangle. Each whole square is 1 square unit.

Online Content at ⇨ **connectED.mcgraw-hill.com**

Try It

Find a formula that can be used to find the area of a rectangle.

1 **Measure the length and width of each object listed in the table.**
Use a centimeter ruler to measure the length and the width of each object to the nearest centimeter. Record the results in the table.

2 **Find the area of each object.**
Use what you learned in the first example to estimate the area of each object. Draw rectangles on grid paper. Then count the unit squares to find the area. Record the results.

Object	Length (cm)	Width (cm)	Area (sq cm)
sticky note			
crayon box			
book			

3 **Justify the formula for area.**
Look for a pattern to find how length and width relate to area.

The area of each object is the product of the
_____ and _____ .

Talk About It

1. How did you make your estimates for the areas of the objects in Activity 2? How close were the estimates to the actual areas?

   _____

   _____

2. What operation can you use with the length and the width to equal the area of a rectangle? Explain.

   _____

   _____

3. **Processes & Practices 2** **Use Symbols** What is the formula for the area of a rectangle? Use A for area, ℓ for length, and w for width.

   _____

840 Chapter 13 Perimeter and Area

Name _____

Practice It

Complete the table below.

Rectangle	Length (units)	Width (units)	Area (sq units)
4.		1	
5.			
6.	6	3	
7.			
8.	7	6	

Lesson 3 Hands On: Model Area 841

Apply It

Use the area formula you wrote in Exercise 3 to solve each problem.

9. Mr. Hart is hanging a picture on a wall. The picture frame has a length of 12 inches and a width of 9 inches. How much wall space will the picture need?

10. What is the area of a classroom with a length of 30 feet and a width of 15 feet?

11. Miss Foster wants to buy carpet for her living room. The living room has a length of 15 feet and a width of 10 feet. How much carpet will she need?

12. **Processes & Practices 5** **Use Math Tools** Ricky's computer monitor is a rectangle. The length is 15 inches and the width is 12 inches. Estimate the area of the monitor.

Write About It

13. Suppose two rectangles have the same area. Must they have the same length and width? Explain.

Name

Lesson 3

Hands On:
Model Area

Homework Helper Need help? connectED.mcgraw-hill.com

Find the area of a parking lot that is 9 yards by 12 yards.

One Way Use a model.

Use grid paper to model the parking lot. Each square equals 1 square yard. Count the number of squares it takes to fill a rectangle that measures 12 units by 9 units.

There are 108 squares, or square units, in all.

Another Way Multiply.

To find the area of a rectangle, multiply the length by the width.

$A = \ell \times w$

$A = 9 \text{ yards} \times 12 \text{ yards}$

$A = 108 \text{ square yards}$

So, the area of the parking lot is 108 square yards.

Practice

Find the area of each rectangle.

1.
   3 units
   8 units

   $A = $ _____ square units

2.
   6 units
   6 units

   $A = $ _____ square units

Lesson 3 My Homework 843

Find the area of each rectangle.

3.

   11 units

   4 units

   A = _____

4.

   7 units

   6 units

   A = _____

Problem Solving

Use the area formula $A = \ell \times w$ you discovered in the lesson to solve each problem.

5. **Processes & Practices 3** **Justify Conclusions** Three guinea pigs living in the same cage need at least 12 square feet of living space. Is a cage that measures 3 feet by 5 feet big enough for three guinea pigs? Explain.

   _____

   _____

6. Jeanette drew a rectangle with an area of 6 square centimeters. What is the possible length and width of her rectangle?

   _____

7. Marcus wants to carpet his clubhouse. One side of the rectangular floor measures 11 feet. Another side measures 8 feet. How many square feet of carpet does Marcus need to completely cover the floor?

   _____

Vocabulary Check

8. Explain the relationship between area and square units.

   _____

9. Define *unit square*.

   _____

Name _____

Lesson 4
Measure Area

ESSENTIAL QUESTION
Why is it important to measure perimeter and area?

You know that area is the number of square units needed to cover a region, or figure, without any overlap.

 Math in My World

5 ft
10 ft

Example 1

The Perez family wants to put the sandbox shown in their backyard. What is the area of the sandbox?

One Way Count unit squares.

Tile the rectangle with unit squares. Each unit square has an area of one square foot.

10 ft
5 ft

There are _____ unit squares.
There are _____ square feet.

So, the area of the sandbox is _____ square feet.

Another Way Multiply.

Multiply the length times the width to find the area.

$A = \text{length} \times \text{width}$

$A = \ell \times w$

$A = 10 \text{ feet} \times 5 \text{ feet}$

$A = $ _____ square feet

Key Concept Area of a Rectangle

Words To find the area A of a rectangle, multiply the length ℓ by the width w.

Symbols $A = \ell \times w$

Online Content at **connectED.mcgraw-hill.com**

Lesson 4 Measure Area **845**

You can also find the area of a square.

Key Concept Area of a Square

Words To find the area A of a square, multiply the length of one side s by itself.

Symbols $A = s \times s$

Example 2

The area and the measure of one side of the square is given. Find the measure of the missing side.

8 m

Area = 64 sq m

$A = s \times s$ Write the formula.

$64 = 8 \times s$

Think: 8 times what number equals 64?

s = _____ meters

The measure of the missing side is _____ meters.

Guided Practice

Find the area of each square or rectangle.

1.   A = _____

Talk MATH

Describe two ways to find the area of a square.

2. 3 yd
   A = _____
    3 yd

846 Chapter 13 Perimeter and Area

Name _____

Independent Practice

Find the area of each square or rectangle.

3.

   A = _____

4.

   A = _____

5.

   A = _____

6.

   A = _____

Algebra The area and the measure of one side of each square or rectangle are given. Label the missing sides.

7.
   7 in.

   Area = 49 square inches

8.
   4 m

   Area = 32 square meters

9.
   1 cm

   Area = 5 square centimeters

10.
    6 in.

    Area = 24 square inches

Lesson 4 Measure Area 847

Problem Solving

11. Each child in Mrs. Dixon's class has a rectangular notebook that has an area of 108 square inches. If the notebook is 9 inches wide, what is the length of the notebook?

___

12. A car is 15 feet long and 6 feet wide. It is parked on a rectangular driveway with an area of 112 square feet. How much of the driveway is not covered by the car?

___

13. Processes & Practices 1 Plan Your Solution A rectangular playground is 40 meters by 10 meters. Its area will be covered with wood chips. Each bag of wood chips covers 200 square meters and costs $30. Find the total cost for this project.

___

Brain Builders

14. Processes & Practices 2 Reason If the sides of a square are doubled, will the area also double? Explain.

___

15. Building on the Essential Question How does finding the area of a rectangle or square with unit squares relate to using the formula to find the area?

___

Name ....

Lesson 4
Measure Area

Homework Helper

Need help? connectED.mcgraw-hill.com

Kristin is making a scrapbook about her pets. She covers one page of the scrapbook with decorative paw print paper. What is the area of this page?

8 inches
8 inches

One Way Count the units.

There are 64 square units in all.

So, the area of the scrapbook page is 64 square inches.

Another Way Multiply.

To find the area of a square, multiply the length of one side by itself.

$A = s \times s$

$A = 8 \text{ inches} \times 8 \text{ inches}$

$A = 64 \text{ square inches}$

Helpful Hint
To find the area of a rectangle, multiply the length by the width.

Practice

Find the area of each figure.

1.

   4 mm
   7 mm

   $A =$ _____ square millimeters

2.

   $A =$ _____ square units

Lesson 4 My Homework 849

Find the area of each figure.

3.

A = _____ square units

4.

5 units
5 units

A = _____ square units

Brain Builders

5. **Processes & Practices 3** Justify Conclusions One side of a square is 10 units. Which is greater, the number of square units for the area of the square or the number of units for the perimeter? Explain.

_____
_____
_____

6. Eric created a rectangular patio using 1-foot square paving stones, which are sold in batches by the dozen. The patio measures 7 feet by 8 feet. How many batches of paving stones did Eric need? Explain. (*Hint:* 1 dozen = 12)

_____
_____

7. **Test Practice** What is the perimeter of the rectangle?

Ⓐ 22 inches Ⓒ 24 inches

Ⓑ 26 inches Ⓓ 28 inches

6 in.
? in.
A = 30 sq in.

Name _____

Lesson 5
Relate Area and Perimeter

ESSENTIAL QUESTION
Why is it important to measure perimeter and area?

Math in My World

Example 1

The swim team put its trophy on a table that has an area of 12 square feet. List all of the possible lengths and widths the table could have.

The models show all of the possible lengths and widths the table could have.

So, the table could have the following possible lengths and widths.

_____ × _____ = 12   _____ × _____ = 12

_____ × _____ = 12   _____ × _____ = 12

_____ × _____ = 12   _____ × _____ = 12

Online Content at connectED.mcgraw-hill.com

Example 2

Find the rectangle with the greatest area whose perimeter is 14 units.

The table shows each rectangle that has a perimeter of 14 units. Complete the table.

Drawing	Rectangle Dimensions	Area
6 by 1 rectangle; $1+6+1+6=14$	1×6	6 square units
5 by 2 rectangle; $2+5+2+5=14$	$2 \times$ ____	____ square units
4 by 3 rectangle; $3+4+3+4=14$	$3 \times$ ____	____ square units

Helpful Hint

The rectangles not listed in the table have dimensions of 6×1, 5×2, and 4×3. If you reverse the dimensions of a rectangle, it will still have the same area.

The greatest area is _____ square units.

So, the rectangle with the greatest area is _____ units by _____ units. Its area is _____ square units.

Talk MATH

Which rectangle in Example 2 has the least area?

Guided Practice

List all the possible dimensions of rectangles for each area.

1. 9 square units

   ____ × ____

   ____ × ____

   ____ × ____

2. 14 square units

   _____

   _____

   _____

Name _____

Independent Practice

List all the possible dimensions of rectangles for each area.

3. 16 square units

4. 20 square units

Find the perimeter and area for each square or rectangle.

5.

6.

Perimeter: _____   Perimeter: _____

Area: _____   Area: _____

7. What do the figures in Exercises 5 and 6 have in common? How do these figures differ?

_____

_____

Lesson 5 Relate Area and Perimeter 853

Problem Solving

8. **Processes & Practices 1** **Plan Your Solution** Violet is making a rectangular banner for the basketball team to run through before the start of the game. She has 24 square feet of paper. List all of the possible dimensions of rectangles with an area of 24 square feet.

9. Which of the dimensions found in Exercise 8 has the greatest perimeter?

Brain Builders

10. If a rectangle has a greater perimeter than another rectangle, does it also have a greater area? Explain. Include a model to support your answer.

11. **Processes & Practices 2** **Reason** Is it possible to draw a rectangle that has an area of 24 square units and a perimeter of 24 units? Explain.

12. **Building on the Essential Question** What is the difference between area and perimeter?

Name _____

MY Homework

Lesson 5
Relate Area and Perimeter

Homework Helper

Need help? connectED.mcgraw-hill.com

A rectangle has a perimeter of 16 units. What is its greatest possible area?

1. Draw all of the possible rectangles with a perimeter of 16 units.

Drawing	Rectangle Dimensions	Area
1, 7, $1+7+1+7=16$	1×7 (or 7×1)	$A = 7$ square units
2, 6, $2+6+2+6=16$	2×6 (or 6×2)	$A = 12$ square units
3, 5, $3+5+3+5=16$	3×5 (or 5×3)	$A = 15$ square units
4, 4, $4+4+4+4=16$	4×4	$A = 16$ square units

If you reverse the dimensions of a rectangle, it will still have the same area.

2. Compare the areas of the rectangles.

The greatest area is 16 square units.

So, 16 square units is the greatest possible area for a rectangle whose perimeter is 16 units.

Lesson 5 My Homework 855

Practice

Draw two possible rectangles for each perimeter. Find the area of each.

1. 20 units

2. 8 units

Brain Builders

3. **Processes & Practices 2** **Use Number Sense** Tomás drew a rectangle with an area of 6 square centimeters. What is the greatest possible perimeter for this rectangle? Explain.

_____

_____

4. Danica has laid out floor tiles so they form a rectangle with a perimeter of 18 inches. What is the difference between the greatest and least possible areas of the rectangle? Explain.

_____

_____

5. A rectangle has an area of 30 square meters and a perimeter of 34 meters. What are the dimensions of the rectangle? Explain.

_____

_____

6. **Test Practice** A square has a perimeter of 28 feet. What is its area?

　Ⓐ 45 square feet　　Ⓒ 49 square feet

　Ⓑ 48 square feet　　Ⓓ 50 square feet

Review

Chapter 13
Perimeter and Area

Vocabulary Check

Use the word bank to complete each sentence.

area   perimeter   square units   unit square

1. The distance around a figure is the _____.

2. _____ is the number of square units needed to cover a region or figure without any overlap.

3. Area is measured in _____.

4. A square with a side length of one unit is called a _____.

Concept Check

Look at the tennis court below. Find the perimeter and area.

5. Perimeter = _____   6. Area = _____

Find each perimeter.

7. 15 cm, 3 cm   P = _____   8. 6 yd, 6 yd   P = _____

My Chapter Review 857

Find each perimeter.

9.

P = _____

10.

P = _____

Find the area of each square or rectangle.

11.

A = _____

12.

A = _____

13.

A = _____

14.

A = _____

15. Find the perimeter and area of the rectangle.

Perimeter: _____

Area: _____

858 Chapter 13 Perimeter and Area

Problem Solving

16. Rodolfo's ping pong table has an area of 45 square feet. The length is 9 feet. What is the perimeter of the ping pong table?

17. Mr. Lobo is building a fence around his rectangular yard. It is 16 feet long and 14 feet wide. How many feet of fencing will he need?

Brain Builders

18. Brett painted the six sides of a cube-shaped box. Each side of the box measures 3 feet. What area did he paint?

19. Is there a relationship between the area and the perimeter of a rectangle? Give an example to help explain your answer.

20. Test Practice Heidi ran two laps around the city block shown. How many feet did she run?

Ⓐ 660 ft Ⓒ 1,320 ft
Ⓑ 880 ft Ⓓ 2,640 ft

Reflect

Chapter 13
Answering the
ESSENTIAL QUESTION

Use what you learned about perimeter and area to complete the graphic organizer.

ESSENTIAL QUESTION
Why is it important to measure perimeter and area?

Perimeter Real-World Examples	Area Real-World Examples

Reflect on the ESSENTIAL QUESTION Write your answer below.

_____
_____
_____

Name _____   Date _____

Score _____

Performance Task

Brain Builders

Home Makeover

Tonya and her family just bought a new house. The house needs a lot of work! The family will fix some things and also build new things in the house to make it their home.

Show all your work to receive full credit.

Part A

The family has to replace the wood floors and baseboards in the dining room. The size of the dining room is shown on the grid. What is the perimeter of the dining room? What is the area? Use formulas to find the answers.

☐ = 1 square unit

Part B

The family needs at least 36 square feet for a new room where they can leave their coats and boots when they come home. They want to use the fewest feet of wall material as possible around the perimeter to save money. Which of these dimensions should they use? Explain.

Length (ft)	Width (ft)	Perimeter (ft)	Area (sq ft)
6	6		
4	9		
5	8		
3	13		

Online Content at connectED.mcgraw-hill.com   Performance Task **860PT1**

Part C

The family has to build a whole new kitchen. They need an area of about 200 square feet so they can fit all the appliances and cabinets. Find two possible dimensions for the new kitchen and show the perimeter and area for each. Then tell which kitchen the family should choose and why.

Kitchen	Length (ft)	Width (ft)	Perimeter (ft)	Area (sq ft)
1				
2				

Part D

The family plans to put down tile in the basement and use the area for a TV room. They decide to put down the tiles in a square, but they are not sure what size the square will be. Complete the table to find what the choices might be.

Length (ft)	Width (ft)	Perimeter (ft)	Area (sq ft)
4		16	
		20	
		24	
		28	

Explain how you completed the table.

860PT2 Chapter 13 Perimeter and Area

Chapter

14 Geometry

Sign Me Up!

ESSENTIAL QUESTION

How are different ideas about geometry connected?

Watch a video!

Name
....

MY Chapter Project

A Banner of Shapes

1. As a group, list 6 shapes on the lines below. Some shapes should be triangles and some should be quadrilaterals.

_____   _____

_____   _____

_____   _____

2. Use a 10-inch square block for each shape. Work with your group to create the shapes. Below each shape, write the name of the shape using fabric paint or glitter glue.

3. Sort your shapes into triangles and quadrilaterals. Draw your sort in the space below.

Name

Am I Ready?

1. Draw three triangles that look different.

Circle the figure described.

2. rectangle

3. square

4. rhombus

5. quadrilateral

Shade the boxes to show the problems you answered correctly.

How Did I Do? | 1 | 2 | 3 | 4 | 5 |

Online Content at connectED.mcgraw-hill.com

Name
....

MY Math Words

Review Vocabulary

rectangle   square   triangle

Making Connections

Match the review vocabulary with each example below. Then describe each shape.

Real-World Object	Name the shape.	Describe the shape.

864　Chapter 14　Geometry

MY Vocabulary Cards

Lesson 14-4

acute angle

Lesson 14-8

acute triangle

Lesson 14-3

angle
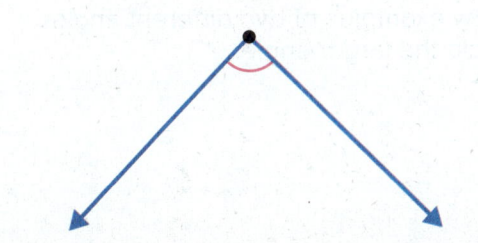

Lesson 14-4

degree (°)

Lesson 14-1

endpoint

Lesson 14-2

intersecting
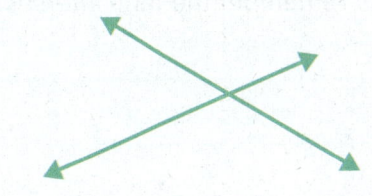

Lesson 14-1

line
line GH
or
$\overleftrightarrow{GH}$

Lesson 14-10

line of symmetry

Ideas for Use

- Draw or write examples for each card. Be sure your examples are different from what is shown on each card.
- Develop categories for the words. Sort them by category. Ask another student to guess each category.

A triangle that has three acute angles.

Does a triangle contain line segments? Explain.

An angle with a measure that is greater than 0° and less than 90°.

Write a tip about the difference between an acute angle and an obtuse angle.

The unit used to measure an angle.

Degree has more than one meaning. Write a definition for a different meaning of the word.

A geometric shape formed when two rays share the same endpoint.

Draw examples of two different angles. Circle the larger angle.

Lines that meet or cross each other.

The Latin root *sect* means "cut." How can this help you remember the term *intersecting*?

The point at the beginning of a ray or line segment.

How many endpoints are on a ray? How many endpoints are on a line segment?

The fold line indicating that a figure's two halves match exactly.

The prefix *a-* can mean "not, without." Draw an asymmetrical object in the space below.

A straight set of points that extend forever in opposite directions.

Line is a multiple-meaning word. Write a sentence using it as a verb.

MY Vocabulary Cards

Processes & Practices

Lesson 14-1

line segment

line segment AB or $\overline{AB}$

Lesson 14-10

line symmetry

Lesson 14-4

obtuse angle

Lesson 14-8

obtuse triangle

Lesson 14-4

one-degree angle

Lesson 14-2

parallel (∥)

Lesson 14-9

parallelogram

Lesson 14-2

perpendicular (⊥)

Ideas for Use

- Design a crossword puzzle. Use the definition for each word as the clues.
- Write a tally mark on each card every time you use the word. Challenge yourself to write at least 3 tally marks on each card.

If a figure can be folded over a line so that one half of the figure matches the other half, it has line symmetry.
How could you demonstrate to a friend that a circle has line symmetry?

A part of a line between two endpoints.
Explain the difference between a line and a line segment.

A triangle with one obtuse angle.
Name one way in which obtuse and acute triangles are similar, and one way in which they are different.

An angle that measures greater than 90° but less than 180°.
Draw an obtuse angle and a right angle below. Circle the obtuse angle.

Lines that are always the same distance apart. Parallel lines do not meet.
Give examples of three shapes that have at least one set of parallel line segments.

An angle that turns through $\frac{1}{360}$ of a circle.
How many one-degree angles are in a circle?

Lines that meet or cross each other to form right angles.
Find two examples of perpendicular line segments in the room.

A quadrilateral with opposites sides equal in length and parallel. Opposite angles have the same size.
Describe the angles of the parallelogram shown on this card.

MY Vocabulary Cards

Lesson 14–1

point

• B

Lesson 14–1

ray

ray MN or $\overrightarrow{MN}$

Lesson 14–9

rectangle

Lesson 14–9

rhombus

Lesson 14–4

right angle

Lesson 14–8

right triangle

Lesson 14–9

square

Lesson 14–9

trapezoid

Ideas for Use

- Practice your penmanship! Write each word in cursive.
- Work with a partner to name a part of speech of each word. Consult a dictionary to check your answers.

A part of a line that has one endpoint and goes on forever in one direction without ending.

What does the multiple-meaning word *ray* mean in this sentence? *A ray of sunshine shone on the wall.*

An exact location represented by a dot.

Write a sentence using the multiple-meaning word *point* as a verb.

A quadrilateral with 4 equal sides. Opposite sides are equal in length and parallel.

Is a rhombus also a trapezoid? Explain.

A quadrilateral with 4 right angles. Both sets of opposite sides are equal in length and parallel.

Solve this riddle: I am a place to play a sport. A net divides each of my sides equally. I'm shaped like a rectangle. What am I?

A triangle with one right angle.

The word *right* has many meanings. Use a dictionary to find a synonym and an antonym for *right*.

An angle with a measure of 90°.

Explain how perpendicular lines and right angles are related.

A quadrilateral with exactly one pair of parallel sides.

Draw a trapezoid and a parallelogram below. Circle the trapezoid.

A quadrilateral with 4 right angles and 4 equal sides. Opposite sides are equal in length and parallel.

Are all squares rectangles? Are all rectangles squares?

Name _____

Lesson 1
Draw Points, Lines, and Rays

ESSENTIAL QUESTION
How are different ideas about geometry connected?

A **point** is an exact location that is represented by a dot. A **line** is a straight set of points that extends in opposite directions without ending.

Math in My World

Example 1

Molly drew the figure shown. Identify the figure she drew.

The figure extends in opposite directions. The arrows indicate that it extends without ending. It is a line.

This line is labeled with point X and point Y. There are different ways to represent this line, such as line XY or $\overleftrightarrow{XY}$.

So, Molly drew _____.

Key Concept Lines, Rays, Line Segments

Words	Models
A **line** is a straight set of points that extends in opposite directions without ending.	line AB or $\overleftrightarrow{AB}$
A **ray** is a part of a line that has one **endpoint** and extends in one direction without ending.	ray AB or $\overrightarrow{AB}$
A **line segment** is a part of a line between two endpoints.	line segment AB or $\overline{AB}$

Online Content at connectED.mcgraw-hill.com

Lesson 1 873

Example 2

Draw a figure that could be represented by $\overline{CD}$.

$\overline{CD}$ represents a line segment with endpoints C and D.

Example 3

Identify the figure at the right.

The figure has one endpoint and extends in one direction without ending. It is a ray.

The endpoint is A. The ray extends in the direction of point B.

So, the figure is _____.

Guided Practice

Identify each figure.

1.

   _____

2.

   _____

3.

   _____

4.

   _____

Talk MATH

How are lines and line segments alike? How are they different?

874 Chapter 14 Geometry

Name

Independent Practice

Identify each figure.

5.
_____

6.
_____

7.
_____

8.
_____

Draw each figure.

9. point T

10. $\overleftrightarrow{YZ}$

11. $\overrightarrow{CR}$

12. $\overline{AW}$

13. $\overleftrightarrow{SN}$

14. $\overrightarrow{TJ}$

Lesson 1 Draw Points, Lines, and Rays 875

Problem Solving

15. Identify the figure that is shown on the stop sign.

_____

16. **Processes &Practices 5** **Use Math Tools** Use a pencil to draw a different type of traffic sign than a stop sign. Then use a crayon or marker to show a line segment on the sign.

Brain Builders

17. **Processes &Practices 4** **Model Math** Name three real-world examples of line segments. How are they different from rays?

_____

_____

18. **Processes &Practices 5** **Use Math Tools** Draw a line segment that is greater than 2 inches and less than 5 inches.

19. **Building on the Essential Question** Explain the similarities and differences of lines and line segments. Draw each figure.

_____

_____

_____

876 Chapter 14 Geometry

Name ....

MY Homework

Lesson 1

Draw Points, Lines, and Rays

Homework Helper

Need help? connectED.mcgraw-hill.com

Abbey is reading the instructions in her drawing book. The instructions are to draw $\overleftrightarrow{AB}$, $\overrightarrow{EK}$, and $\overline{JT}$. What figures should she draw?

A line is a straight set of points that extend in opposite directions without ending. You can represent it as line AB or $\overleftrightarrow{AB}$.

A ray is a figure that has one endpoint and extends in one direction without ending. You can represent it as ray EK or $\overrightarrow{EK}$.

A line segment is a part of a line between two endpoints. You can represent it as line segment JT or $\overline{JT}$.

Practice

Identify each figure.

1.

2.

Lesson 1 My Homework 877

Draw each figure.

3. $\overline{YB}$

4. $\vec{LR}$

Brain Builders

5. Create a drawing that includes 3 line segments, 1 ray, and 4 points.

Vocabulary Check

Choose the correct word(s) to complete each sentence.

    endpoint   line   line segment   point   ray

6. A _____ is part of a line between two endpoints.

7. An exact location represented by a dot is a _____.

8. A _____ is a figure that extends in opposite directions without ending.

9. A ray has one _____.

10. A _____ extends in one direction without ending.

11. **Test Practice** Which is the correct way to represent the figure?

    Ⓐ ray WB   Ⓒ $\overline{WB}$

    Ⓑ line segment WB   Ⓓ $\overleftrightarrow{WB}$

878   **Need more practice?** Download Extra Practice at connectED.mcgraw-hill.com

Name _____

Lesson 2
Draw Parallel and Perpendicular Lines

ESSENTIAL QUESTION

How are different ideas about geometry connected?

You can describe lines, rays, and line segments by the way they cross each other or do not cross each other.

 Math in My World

Example 1

Oliver was riding in the car and saw this sign. Describe how the outlined line segments cross each other or do not cross each other.

Parallel lines are always the same distance apart. They do not meet or cross each other.

So, Oliver saw a figure with _____ line segments.

Key Concept Types of Lines

Words Parallel lines are always the same distance apart. They do not meet.

Model

line AB is parallel to line CD

$\overleftrightarrow{AB} \parallel \overleftrightarrow{CB}$

The symbol || means parallel.

Online Content at connectED.mcgraw-hill.com

Key Concept — Types of Lines

Words Lines that meet or cross each other are called **intersecting** lines.

Model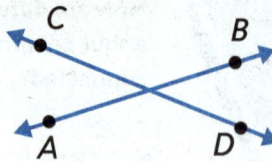

line AB intersects line CD
$\overleftrightarrow{AB}$ intersects $\overleftrightarrow{CD}$

Words Lines that meet or cross each other to form square corners are called **perpendicular** lines.

Model

line AB is perpendicular to line CD
$\overleftrightarrow{AB} \perp \overleftrightarrow{CD}$

The symbol ⊥ means perpendicular.

The symbol ⌐ means square corner.

Example 2

Describe the figure. Use *parallel*, *perpendicular*, or *intersecting*. Use the most specific term.

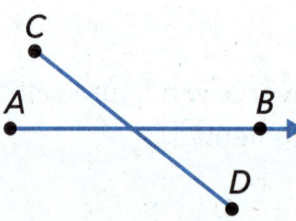

The figure shows ray AB and line segment CD.

The figures cross each other but do not form square corners.

$\overrightarrow{AB}$ and $\overline{CD}$ are _____.

Guided Practice

1. Describe the line segments outlined on the tennis racquet.

Talk MATH
Name a real-world example of parallel line segments and intersecting line segments.

880 Chapter 14 Geometry

Name

Independent Practice

Describe each figure. Use *parallel, perpendicular,* or *intersecting*.
Use the most specific term.

2.

$\overleftrightarrow{LM}$ and $\overrightarrow{JK}$ are

_____

3.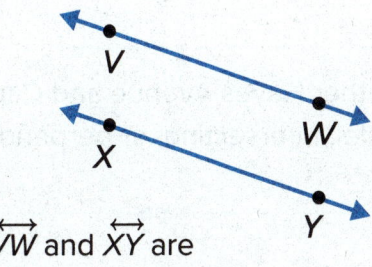

$\overleftrightarrow{VW}$ and $\overleftrightarrow{XY}$ are

_____

Draw an example of each figure.

4. $\overleftrightarrow{DE} \parallel \overleftrightarrow{FG}$

5. $\overleftrightarrow{RS}$ intersects $\overleftrightarrow{TU}$

6. $\overleftrightarrow{NO} \perp \overleftrightarrow{PQ}$

7. $\overline{JK} \parallel \overrightarrow{LM}$

8. Circle the statement that is true about the figure below.

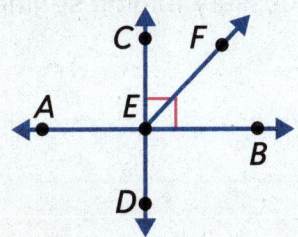

Line *AB* is parallel to ray *EF*.

Line *AB* is perpendicular to line *CD*.

Line *CD* is parallel to ray *EF*.

Line *CD* is parallel to line *AB*.

Lesson 2 Draw Parallel and Perpendicular Lines **881**

Problem Solving

Processes & Practices 4 **Model Math** On a map, streets can be represented by line segments. Use the map to answer Exercises 9–11.

9. Identify two streets that appear to be parallel to Washington Avenue.

   _____

10. Tell whether Hayes Avenue and Capital Lane appear to be parallel, intersecting, or perpendicular lines. Explain.

    _____

    _____

11. Are there any streets that are intersecting but not perpendicular? Explain.

    _____

    _____

    _____

Brain Builders

12. **Processes & Practices 2** **Stop and Reflect** Tell whether each statement is *true* or *false*. If *false*, correct the statement so that it is true.

    - If two lines are parallel, they are always the same distance apart. _____

    - If two lines are parallel, they are also perpendicular. _____

13. **Building on the Essential Question** Describe a real-world example of when it is necessary for line segments to be parallel and when it is necessary for line segments to be perpendicular.

    _____

    _____

882 Chapter 14 Geometry

Name

MY Homework

Lesson 2

Draw Parallel and Perpendicular Lines

Homework Helper

Need help? connectED.mcgraw-hill.com

Describe each figure. Use *parallel*, *perpendicular*, or *intersecting*. Use the most specific term.

Lines or line segments that are always the same distance apart and do not meet are parallel. So, $\overline{RK} \parallel \overline{DM}$.

Lines or line segments that meet to form square corners are perpendicular. So, $\overleftrightarrow{PB} \perp \overleftrightarrow{EH}$.

Lines or line segments that meet or cross are intersecting. So, $\overleftrightarrow{XY}$ intersects $\overleftrightarrow{CF}$.

Helpful Hint
The symbol ∥ means parallel.
The symbol ⊥ means perpendicular.

Practice

1. Describe the figure. Use *parallel*, *perpendicular*, or *intersecting*. Use the most specific term.

_____

Lesson 2 My Homework 883

Processes & Practices **2** Use Symbols Draw an example of each figure.

2. $\overleftrightarrow{GP} \parallel \overrightarrow{ND}$

3. $\overrightarrow{HY}$ intersects $\overline{QA}$

Brain Builders

4. **Processes & Practices** **4** Model Math Martin incorrectly claims that there are only parallel line segments found in the blinds. Give an example of perpendicular line segments found in the blinds.

_____

_____

Vocabulary Check

Draw a line to match each vocabulary term to its example.

5. intersecting, but not perpendicular

6. parallel

7. perpendicular

8. **Test Practice** Which is the correct description of the figure?

Ⓐ $\overleftrightarrow{QB} \parallel \overleftrightarrow{SF}$

Ⓒ $\overleftrightarrow{QS} \perp \overleftrightarrow{BF}$

Ⓑ $\overleftrightarrow{QB} \perp \overleftrightarrow{SF}$

Ⓓ $\overleftrightarrow{QS} \parallel \overleftrightarrow{BF}$

Check My Progress

Vocabulary Check

Use the word bank to complete each sentence.

endpoint   line   line segment   point   ray

1. A _____ is part of a line between two endpoints.

2. A _____ is a part of a line that has one _____ and extends in one direction without ending.

3. A _____ is a straight set of points that extends in opposite directions without ending.

4. A _____ is an exact location that is represented by a dot.

Match each vocabulary word to its definition.

5. **intersecting** • lines that meet or cross each other to form square corners

6. **parallel** • lines that meet or cross each other, but do not necessarily form square corners

7. **perpendicular** • lines that are always the same distance apart and do not meet

Concept Check

Circle the correct description of each figure.

8. M—N

   line
   line segment
   ray

9. P—O

   line
   line segment
   ray

10. Q—R

    line
    line segment
    ray

Problem Solving

11. Identify the lines outlined on the pair of scissors as parallel, perpendicular, or intersecting. Choose the most specific term.

   _____

12. Sandy is driving on Broadway Avenue. Which street appears to be perpendicular to Broadway Avenue?

   _____

Brain Builders

13. Nathan practiced his handwriting by writing the alphabet in capital letters. He wrote all of the letters that have both parallel and perpendicular line segments. Which letters did Nathan write?

   _____

14. **Test Practice** Which figure shows parallel lines?

   Ⓐ   Ⓒ

   Ⓑ   Ⓓ

Name

Lesson 3
Hands On
Model Angles

ESSENTIAL QUESTION
How are different ideas about geometry connected?

An **angle** is a geometric shape that is formed when two rays have the same endpoint.

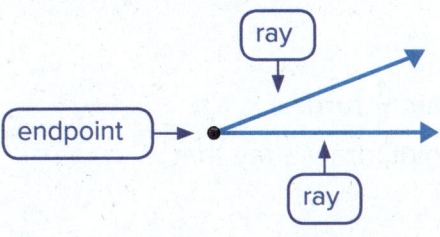

Angles are measured by the amount of rotation, or turning, from one ray to another.

$\frac{1}{4}$ turn $\frac{1}{2}$ turn full turn

Draw It

Trace the lines to draw an angle that measures $\frac{1}{4}$ turn.
Then draw an angle that measures less than $\frac{1}{4}$ turn.

 Draw an angle that measures $\frac{1}{4}$ turn.

Trace the lines to draw two rays with a common endpoint. The two rays form an angle.

The center of the circle is at the same point as the endpoint of the two rays. The angle you drew measures $\frac{1}{4}$ turn.

 Draw an angle that measures less than $\frac{1}{4}$ turn.

Using the center of the circle as an endpoint, draw a ray that is inside the angle you drew in Step 1.

The ray you drew and the horizontal ray form an angle. This angle measures less than $\frac{1}{4}$ turn.

Online Content at connectED.mcgraw-hill.com

Try It

Trace the lines to draw an angle that measures $\frac{1}{4}$ turn.
Then draw an angle that measures greater than $\frac{1}{4}$ turn.

 Draw an angle that measures $\frac{1}{4}$ turn.

Trace the lines to draw two rays with a common endpoint. The two rays form an angle.

The center of the circle is at the same point as the endpoint of the two rays.

The angle you drew measures $\frac{1}{4}$ turn.

 Draw an angle that measures greater than $\frac{1}{4}$ turn.

Using the center of the circle as an endpoint, draw a ray that is outside the angle you drew in Step 1.

The ray you drew and the horizontal ray form an angle.

This angle measures greater than $\frac{1}{4}$ turn.

Talk About It

1. **Draw a Conclusion** Without drawing a circle, determine whether the angle at the right has a measure that is greater than, less than, or equal to $\frac{1}{2}$ turn. Explain.

_____

_____

2. **Processes &Practices** 6 **Explain to a Friend** Refer to the angle you drew in the Try It activity above. State whether the angle's measure is greater than, less than, or equal to $\frac{1}{2}$ turn. Explain.

_____

_____

_____

Name ....

Practice It

3. Draw an angle with a measure less than $\frac{1}{4}$ turn.

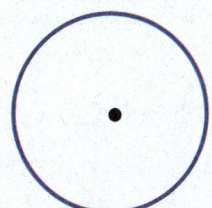

4. Draw an angle with a measure greater than $\frac{1}{4}$ turn.

Draw lines to match each figure to its description.

5.

 An angle with a measure greater than $\frac{1}{4}$ turn, but less than $\frac{1}{2}$ turn.

6.

 An angle with a measure less than $\frac{1}{4}$ turn.

7.

 An angle with a measure of $\frac{1}{4}$ turn.

8.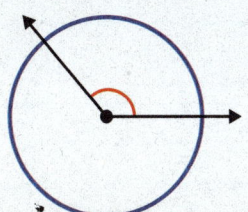

• An angle with a measure of $\frac{1}{2}$ turn.

Lesson 3 Hands On: Model Angles 889

Apply It

9. Draw the hands on the clock below to show 5:00.

Describe the measure of the angle formed by the hands of the clock.

_____

_____

_____

10. Draw a real-world object that shows an angle with a measure of $\frac{1}{4}$ turn.

11. Processes & Practices   **4**   **Model Math** Draw two angles that share a common endpoint and a common ray. Together, they should form an angle with a measure of $\frac{1}{4}$ turn.

Write About It

12. How can I describe an angle's measure?

_____

_____

_____

890   **Chapter 14** Geometry

Name ....

MY Homework

Lesson 3

Hands On: Model Angles

Homework Helper

Need help? connectED.mcgraw-hill.com

Draw an angle with a measure greater than $\frac{1}{4}$ turn.

Draw two rays with a common endpoint at the center of the circle.
Be sure that the angle created by the rays has a measure greater than $\frac{1}{4}$ turn.

How do you know the measure of the angle is greater than $\frac{1}{4}$ turn?

An angle with a measure of $\frac{1}{4}$ turn forms a square corner.
The angle drawn has a greater measure than $\frac{1}{4}$ turn.

Practice

Circle the correct description for the measure of each angle.

1.

less than $\frac{1}{4}$ turn

greater than $\frac{1}{4}$ turn

2.

less than $\frac{1}{4}$ turn

greater than $\frac{1}{4}$ turn

3.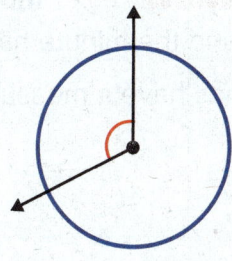

less than $\frac{1}{2}$ turn

greater than $\frac{1}{2}$ turn

Lesson 3 My Homework **891**

Draw an angle that has each measure.

4. $\frac{1}{4}$ turn

5. $\frac{1}{2}$ turn

6. greater than $\frac{1}{2}$ turn

7. less than $\frac{1}{4}$ turn

Problem Solving

8. **Processes & Practices 6** **Be Precise** Describe the rays of an angle that has a measure of $\frac{1}{2}$ turn.

9. **Processes & Practices 4** **Model Math** Imagine a clock with the hour hand at 12 and the minute hand at 2. Does the angle formed by the two hands have a measure greater than, less than, or equal to $\frac{1}{4}$ turn?

Vocabulary Check

10. In your own words, define *angle*.

892 Chapter 14 Geometry

Name _____

Lesson 4
Classify Angles

ESSENTIAL QUESTION
How are different ideas about geometry connected?

Angles can be measured in a more precise way than turns. The unit used to measure an angle is called a **degree (°)**. A circle is made up of 360°.

An angle that turns through $\frac{1}{360}$ of a circle is called a **one-degree angle**. That means that 360 one-degree angles sharing the same endpoint make a circle. The angle below turns through 3 one-degree angles. So, it measures 3°.

Math in My World

Example 1

David walks by a crosswalk sign on his way to school. The angle outlined on the sign turns through 60 one-degree angles. Find the measure of the angle.

The angle turns through 60 one-degree angles.

That means that 60 one-degree angles sharing the same endpoint make the angle.

So, the angle has a measure of _____°.

Online Content at connectED.mcgraw-hill.com

Angles can be classified as *right*, *acute*, or *obtuse*.

Key Concept — Types of Angles

A **right angle** measures 90°.

This symbol means right angle.

An **acute angle** measures greater than 0° and less than 90°.

An **obtuse angle** measures greater than 90° but less than 180°.

Example 2

Classify the angle as *right*, *acute*, or *obtuse*.
The angle is 90°.

So, it is a(n) _____ angle.

Example 3

Classify the angle as *right*, *acute*, or *obtuse*.
The angle is greater than 90° and less than 180°.

So, it is a(n) _____ angle.

Guided Practice

1. The angle shown turns through 94 one-degree angles. Find the measure of the angle.

   _____

2. Classify the angle shown as *right*, *acute*, or *obtuse*.

   _____

Talk MATH

How many one-degree angles does a right angle turn through?

894 **Chapter 14** Geometry

Name _____

Independent Practice

Write the measure of the angle in degrees and as a fraction of a full turn.

3.

   _____

4.

   _____

Classify each angle as *right*, *acute*, or *obtuse*.

5.

   _____

6.

   _____

7.

   _____

8.

   _____

Draw an example of each figure.

9. an acute angle

10. an obtuse angle

11. An angle that measures 30° turns through how many one-degree angles? _____

12. Classify the angle in Excercise 11 as *acute*, *right*, or *obtuse*. _____

13. An angle that measures 100° turns through how many one-degree angles? _____

14. Classify the angle in Excercise 13 as *acute*, *right*, or *obtuse*. _____

Lesson 4 Classify Angles 895

Problem Solving

15. The timer is set to 30 minutes. Through how many degrees will the dial have turned when the timer goes off? What fraction of a full turn is shown by the angle?

_____

16. Classify the angle shown on the gas gauge.

_____

Brain Builders

17. Processes &Practices 6 **Be Precise** Draw three angles that satisfy the clues below.

- The first angle is a right angle.
- The second angle turns through more one-degree angles than the first angle.
- The third angle turns through less one-degree angles than the first angle.

Classify the second and third angles as *acute, right,* or *obtuse*.

_____

18. Building on the Essential Question How is a one-degree angle helpful in classifying angles? Include a drawing to support your answer.

_____

_____

_____

896 Chapter 14 Geometry

Name _____

MY Homework

Lesson 4
Classify Angles

Homework Helper

Need help? connectED.mcgraw-hill.com

Classify the angle on the inside of the goalpost as *right*, *acute*, or *obtuse*.

The angle on the inside of the goalpost forms a square and measures 90°.

So, it is a right angle.

Practice

Write the measure of each angle in degrees and as a fraction of a full turn.

1.

   _____°, _____ turn

2.

   _____°, _____ turn

3.

   _____°, _____ turn

4.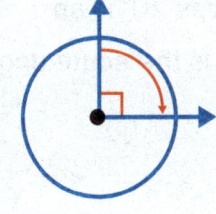

   _____°, _____ turn

Lesson 4 My Homework 897

Classify each angle as *right*, *acute*, or *obtuse*.

5. _____

6. _____

Brain Builders

7. **Processes & Practices 4** **Model Math** The hour and minute hands of a clock form an angle that constantly changes. During each hour of the day, the clock hands will form right, acute and obtuse angles. Assume the hour hand points directly at the hour. Write a time of day when the two clock hands form each type of angle.

   _____

   _____

8. Amelia starts reading at 3:25 and reads for 15 minutes. What type of angle is formed by the hour and minute hands of the clock when Amelia finishes reading?

   _____

Vocabulary Check

Write a vocabulary term to complete each sentence.

   acute angle   degree   obtuse angle   right angle

9. The unit used to measure an angle is a _____.

10. An _____ measures more than 90° and less than 180°.

11. *Square corner* is another name for a _____.

12. An angle that measures 20° is an _____.

13. **Test Practice** Which is the acute angle?

   Ⓐ   Ⓒ

   Ⓑ   Ⓓ

Name ....

Lesson 5
Measure Angles

ESSENTIAL QUESTION
How are different ideas about geometry connected?

The protractor is a tool used to measure angles. The length of the rays does not affect the measure of the angle.

Math in My World

Example 1

Henry drew the angle shown. Measure the angle.

1. **Line up the protractor.**

   Place the center of the protractor on the endpoint of the angle with the straightedge along one ray.

2. **Line up the angle.**

   Line up one ray of the angle with the zero on the protractor.

3. **Measure the angle.**

   Find the tic mark on the protractor that aligns with the second ray of the angle.

   So, the measure of the angle is _____.

Online Content at connectED.mcgraw-hill.com

Example 2

Measure the angle.

Use a protractor.

 Line up the protractor.

 Line up the angle.

 Measure the angle.

So, the angle measures _____.

Check Be sure that you read the protractor correctly. The angle is obtuse, so choose 125°, not 55°, as the measure of the angle.

Guided Practice

Use a protractor to measure each angle.

1.
_____

2.
_____

3. Measure the indicated angle on the triangle.

_____

Talk MATH

Explain how to use a protractor.

Name

Independent Practice

Use a protractor to measure each indicated angle.

4.

_____

5.

_____

6.

_____

7.

_____

8.

_____

9.

_____

10.

_____

11.

_____

12. Refer to Exercises 8–11. Which exercises show figures with at least one acute angle?

_____

Lesson 5 Measure Angles 901

Problem Solving

13. Darius has a square picture frame. He measures the angle of one corner. Is the angle 90° or 145°? Explain your reasoning.

_____

_____

14. **Processes & Practices 5** **Use Math Tools** Sue drew two angles that share a common endpoint and a common ray. One angle is 20°. Measure the other angle. Then measure the angle formed by the two angles together.

_____

Brain Builders

15. The angles shown have a total measure of 170°. One angle is 40°. Use a protractor to find the measure of the other angle. Write an equation to show the total measurement of the two angles.

_____

16. **Processes & Practices 5** **Use Math Tools** Draw two different quadrilaterals that each have at least one angle that measures less than 90°. What do you notice about the measures of the other angles?

_____

17. **Building on the Essential Question** Why is it important to line up a protractor correctly when measuring an angle?

_____

_____

Name ....

Lesson 5
Measure Angles

Homework Helper

Need help? connectED.mcgraw-hill.com

Camille is making the pinwheel shown below. Measure the angle shown in red.

Use a protractor.

 Line up the protractor.

 Line up the angle.

 Measure the angle.

So, the angle measures 100°.

Practice

Use a protractor to measure each angle.

1.

2.

3.

4.

Use a protractor to measure each angle outlined in red.

5.

6.

_____   _____

7.

8.

_____   _____

Brain Builders

9. Heong cut a slice of birthday cake. The slice formed the angle shown. Natalie cut a slice of birthday cake with an angle that measures 40°. Who cut the larger slice of cake?

_____

10. **Processes & Practices 5** Use Math Tools Dimitri drew a right angle. Then he drew an angle that was 20° larger. What is the measure of the second angle Dimitri drew?

_____

11. **Test Practice** What is the measure of the angle?

Ⓐ 85°   Ⓒ 75°

Ⓑ 80°   Ⓓ 70°

Name _____

Lesson 6
Draw Angles

ESSENTIAL QUESTION
How are different ideas about geometry connected?

You have used a protractor to measure angles. A protractor can also be used to draw angles of a given measure.

 Math in My World

Example 1
A traffic sign shows an 80° angle. Draw an 80° angle.

 Draw one ray of the angle.
Mark the endpoint and draw a ray.

 Measure the angle.
Place the protractor along the ray as you would to measure an angle. On the protractor, find 80°, and make a pencil mark.

 Draw the next ray of the angle.
Use a ruler to draw the ray that connects the endpoint and the pencil mark.

Draw your 80° angle below.

Online Content at connectED.mcgraw-hill.com

Lesson 6 905

Example 2

Draw an angle that is greater than 10° and less than 30°. Measure and classify the angle.

Find 10° and 30° on the protractor. Draw an angle between the two measurements. Then measure the angle drawn.

So, the angle drawn is _____. It is a(n) _____ angle.

Guided Practice

Draw an angle for each measure.

1. 20°

2. 45°

3. 100°

Talk MATH

Describe how you would draw a 90° angle without using a protractor.

906 Chapter 14 Geometry

Name: ....

Independent Practice

Draw an angle for each measure.

4. 10°

5. 75°

6. 90°

7. 115°

8. 140°

9. 135°

Draw an angle between the given measurements. Measure and classify each angle as *acute*, *right*, or *obtuse*.

10. 0° and 20°

11. 90° and 120°

angle measure: _____

angle measure: _____

type of angle: _____

type of angle: _____

Lesson 6 Draw Angles   **907**

Problem Solving

12. **Processes &Practices 5** **Use Math Tools** Julia drew an angle that is greater than 45° and less than 100°. Draw an angle that could be the angle that Julia drew. Then draw an angle that could not be the angle that Julia drew.

Explain why the angle cannot be Julia's angle.

_____

Brain Builders

13. **Processes &Practices 3** **Find the Error** Ethan was asked to draw a 130° angle. He drew the angle below. Find and correct his mistake.

_____

_____

14. **Building on the Essential Question** How can I check the reasonableness of my angle measurements?

_____

_____

_____

Name _____

Lesson 6

Draw Angles

Homework Helper

Need help? connectED.mcgraw-hill.com

Draw a 40° angle. Classify it as *acute*, *right*, or *obtuse*.

1 **Draw one ray of the angle.**
Mark the endpoint and draw a ray.

endpoint

2 **Measure the angle.**
Place the protractor along the ray as you would to measure an angle. On the protractor, find 40°, and make a pencil mark.

3 **Draw the next ray of the angle.**
Use a ruler to draw the ray that connects the vertex and the pencil mark.

The angle is less than 90°, so it is an acute angle.

Practice

Draw an angle for each measure.

1. 65°

2. 140°

Lesson 6 My Homework 909

Draw an angle for each measure.

3. 80°

4. 35°

5. greater than 5° and less than 25°

6. greater than 90° and less than 120°

Brain Builders

7. Processes & Practices 6   Be Precise What type of angle is formed by joining an acute angle and a right angle? Explain.

_____

_____

8. Lila draws a 145° angle. Then she divides it into three smaller angles. One of the smaller angles measures 65°, and the other two are equal in measure. What is the measure of the other two angles? Explain.

_____

9. Test Practice Which is the correct drawing of a 160° angle?

910 **Need more practice?** Download Extra Practice at connectED.mcgraw-hill.com

Name _____

Lesson 7
Solve Problems with Angles

ESSENTIAL QUESTION
How are different ideas about geometry connected?

An angle can be decomposed, or broken, into non-overlapping parts. The angle measure of the whole is the sum of the angle measures of the parts.

Math in My World

Example 1
Rachel and Dean made a sign out of fabric like the one shown to hang in the school gymnasium. The blue piece has a 35° angle. The red piece is attached to the longest side of the blue piece. Together, the pieces form a right angle. What is the angle shown on the red piece?

One Way Make a model.

Draw a 90° angle. Mark off a 35° angle. Measure the other angle.

The other angle has a measure of _____.

Another Way Use an equation.

The 90° angle measure is the sum of two parts. One angle is 35°. Find the unknown angle measure.
Let r represent the unknown angle measure.

$35 + r = 90$

Since $35 + r = 90$, you know that $90 - 35 = r$.

$r = 90 - 35$

$r =$ _____

So, the angle shown on the red piece measures _____.

Addition and subtraction are inverse, or opposite, operations.

Online Content at connectED.mcgraw-hill.com

Lesson 7   911

Example 2

Find the combined measure of the angle shown.

One of the angles is 20°. The symbol on the other angle shows that it is a right angle. Therefore, it is 90°.

To find the combined measure of the angle, add the angle measures of the parts.

Let a represent the combined angle measure.

$a = 20° + 90°$

$a = $ _____

So, the combined measure of the angle is _____.

Guided Practice

Algebra Find each unknown.

1. The combined angle measure is 90°.

$n = $ _____

2. The combined angle measure is 130°.

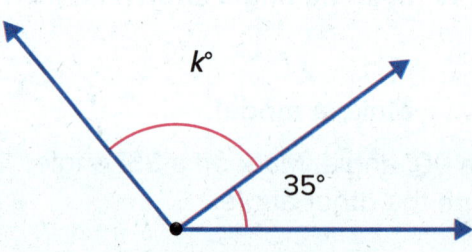

$k = $ _____

3. Find the combined angle measure.

combined measure = _____

Talk MATH

How can the measures of parts of an angle be used to find the combined measure?

912 Chapter 14 Geometry

Name ....

Independent Practice

Algebra Find each unknown.

4. The combined angle measure is 50°.

a = _____

5. The combined angle measure is 90°.

n = _____

6. The combined angle measure is 125°.

g = _____

7. The combined angle measure is 150°.

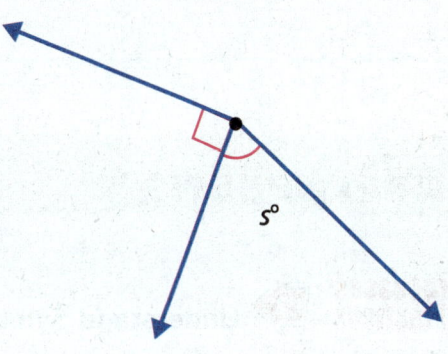

s = _____

8. Draw a triangle with one right angle.

Find the combined measure of the three angles.

_____

9. Draw a triangle with one obtuse angle.

Find the combined measure of the three angles.

_____

Lesson 7 Solve Problems with Angles 913

Problem Solving

10. The steps on a staircase should be 90°. One of the steps is crooked. The angle formed is 15° too large. What is the angle of that step?

_____

11. Processes & Practices 4 **Model Math** The combined measure of the angles is 150°. One angle measures 50°. Find the value of x.

$x =$ _____

Brain Builders

12. Processes & Practices 2 **Understand Symbols** Find the value of k. Explain how you solved the problem.

$k =$ _____

_____

13. Building on the Essential Question How is addition related to angle measurement? Include a drawing to support your answer.

_____

_____

_____

914 Chapter 14 Geometry

Name

MY Homework

Lesson 7

Solve Problems with Angles

Homework Helper

Need help? connectED.mcgraw-hill.com

Find the measure of the unknown angle. The combined angle measure is 140°.

Use an equation.

You know the total measure is 140°. One angle is 25°.

Let d represent the unknown angle measure.

$$25 + d = 140$$

Since $25 + d = 140$, you know that $140 - 25 = d$. ← Addition and subtraction are inverse, or opposite, operations.

$$d = 140 - 25$$

$$d = 115$$

So, the measure of the unknown angle is 115°.

Practice

Algebra Find each unknown.

1. The combined angle measure is 50°.

$h = $ _____

2. The combined angle measure is 135°.

$p = $ _____

Lesson 7 My Homework 915

Algebra Find each unknown.

3. The combined angle measure is 70°.

f = _____

4. The combined angle measure is 115°.

t = _____

5. The combined angle measure is 180°.

150°, x°

x = _____

6. Find the value of r.

r = _____

Brain Builders

7. **Processes & Practices 1** **Make a Plan** Suppose you draw a line from the center of a clock face to the number 12. When the minute hand gets to 3 on the clock face, the line and minute hand form a 90° angle. What angle do the line and the minute hand make when the minute hand is on 2? Explain.

_____

_____

8. **Test Practice** The combined angle measure is 120°. What is the value of $(n + 20)°$?

Ⓐ 20

Ⓑ 40

Ⓒ 60

Ⓓ 80

916 **Need more practice?** Download Extra Practice at connectED.mcgraw-hill.com

Check My Progress

Vocabulary Check

Unscramble each word to complete each sentence. Use a word from the word bank.

acute angle   **angle**   **degree (°)**   **obtuse angle**   **right angle**

1. thrig lngae

    A _____ measures 90°.

2. leagn

    An _____ is a geometric shape that is formed when two rays have the same endpoint.

3. tacue aleng

    An _____ measures greater than 0° and less than 90°.

4. geeedr

    The unit used to measure an angle is called a _____.

5. oseutb egnla

    An _____ measures greater than 90° but less than 180°.

Concept Check

Classify each angle as *right*, *acute*, or *obtuse*.

6.

7.

8.

Problem Solving

9. An angle on a grocery cart is made up of 100 one-degree angles. Find the measure of the angle.

10. The angle shown is made up of 112 one-degree angles. Find the measure of the angle.

Brain Builders

11. Draw two angles whose measures add to 85°. What type of angles are the two angles?

12. **Test Practice** What is the measure of the angle in degrees and as a fraction of a full turn?

    Ⓐ 90°; $\frac{3}{4}$ turn

    Ⓑ 90°; $\frac{1}{360}$ turn

    Ⓒ 90°; $\frac{1}{4}$ turn

    Ⓓ 180°; $\frac{1}{2}$ turn

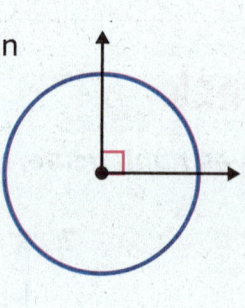

918 Chapter 14 Geometry

Name _____

Lesson 8
Triangles

ESSENTIAL QUESTION
How are different ideas about geometry connected?

There are many different kinds of triangles. You can classify triangles by the measure of their angles.

 Math in My World

Lunch time!

Example 1

This sandwich is cut in half. Classify the triangle represented by the half sandwich as *right*, *acute*, or *obtuse*. Determine if any of the sides are perpendicular.

A **right triangle** has one right angle.

How many right angles are there? _____

The two sides that form the right angle are perpendicular.

An **acute triangle** has three acute angles.

How many acute angles are there? _____

An **obtuse triangle** has one obtuse angle.

How many obtuse angles are there? _____

So, the triangle is a(n) _____ triangle.

Key Concept   Classify Triangles by Angles

An obtuse triangle has one obtuse angle.

An acute triangle has three acute angles.

A right triangle has one right angle. The two sides that form the right angle are perpendicular.

Online Content at connectED.mcgraw-hill.com

Lesson 8 919

You can also identify vertices and line segments in triangles.

Key Concept Vertices and Line Segments in Triangles

A triangle has three vertices and three line segments.

Each point is formed by the intersection of two line segments.

Vertices: A, B, C

Line Segments: $\overline{AB}$, $\overline{BC}$ and $\overline{AC}$

Example 2

Classify the triangle as *right*, *acute*, or *obtuse*. Identify the vertices and line segments of the triangle.

The triangle is a(n) _____ triangle because it has one _____ angle.

There are _____ vertices. List them below.

Vertices: _____

There are _____ line segments. List them below.

Line Segments: _____

Talk MATH
Is it possible for a triangle to have a pair of parallel sides? Explain.

Guided Practice

1. Classify the triangle as *acute*, *right*, or *obtuse*. Determine how many sides are perpendicular.

The triangle is _____.

_____ sides are perpendicular.

920 Chapter 14 Geometry

Name ....

Independent Practice

Classify each triangle as *acute*, *right*, or *obtuse*.
Circle the triangles that have any perpendicular sides.

2.

3.

4.

_____   _____   _____

5.

6.

7.

_____   _____   _____

8. Draw three line segments that form a right triangle.

9. Draw three line segments that form an obtuse triangle.

10. Which exercises on this page show right triangles?
_____

11. Which exercises on this page show figures with perpendicular line segments?
_____

Lesson 8 Triangles 921

Problem Solving

For Exercises 12 and 13, refer to the map of Arizona at the right.

12. Classify the triangle formed by Phoenix, Yuma, and the Grand Canyon as *acute*, *right*, or *obtuse*.

_____

13. **Processes & Practices 6** **Explain to a Friend** Explain how you classified the triangle.

_____

_____

Brain Builders

14. Ved measured the angles of a triangle as 25°, 75°, and 80°. Is this triangle acute, right, or obtuse? Explain.

_____

15. **Processes & Practices 3** **Find the Error** Ben said that the triangle shown is an acute triangle because the angle shown is acute. Find and correct his mistake.

_____

_____

16. **?** **Building on the Essential Question** Is it possible for a triangle to have two obtuse angles? Explain.

_____

_____

_____

Chapter 14 Geometry

Name ....

MY Homework

Lesson 8

Triangles

Homework Helper

Need help? connectED.mcgraw-hill.com

Identify the vertices and line segments of each triangle. Identify each triangle as *right*, *acute*, or *obtuse*.

There are three vertices. They are J, M, and P.

There are three line segments. They are $\overline{JM}$, $\overline{MP}$, and $\overline{JP}$.

The triangle has 3 acute angles, so it is an acute triangle.

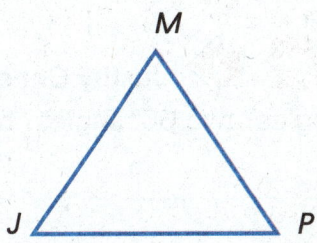

The vertices are A, B, and C.

The three line segments are $\overline{AB}$, $\overline{BC}$, and $\overline{AC}$.

The triangle has a right angle, so it is a right triangle.

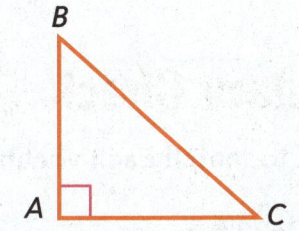

The three vertices are T, R, and G.

The three line segments are $\overline{TR}$, $\overline{RG}$, and $\overline{TG}$.

The triangle has an obtuse angle, so it is an obtuse triangle.

Practice

Identify the perpendicular sides of each right triangle.

1.

2.

Lesson 8 My Homework 923

Classify each triangle as *acute*, *right*, or *obtuse*.
Determine how many sides appear to be perpendicular.

3.

4.

_____   _____

Brain Builders

5. **Processes & Practices 3** **Justify Conclusions** Can a triangle have two 30° angles? two 60° angles? two 90° angles? Explain.

_____

_____

Vocabulary Check

Draw a line to match each vocabulary term to its example.

6. acute triangle

7. obtuse triangle

8. right triangle

9. **Test Practice** Which is an acute triangle?

Ⓐ   Ⓑ   Ⓒ   Ⓓ

Name _____

Lesson 9
Quadrilaterals

ESSENTIAL QUESTION
How are different ideas about geometry connected?

All quadrilaterals have 4 sides and 4 angles. There are many different kinds of quadrilaterals.

Math in My World

Example 1

The speed limit sign represents a quadrilateral. Classify the angles formed by the quadrilateral. Determine if any of the sides are parallel or perpendicular.

1. Classify the angles.

There are _____ right angles,

_____ acute angles, and _____ obtuse angles.

2. Determine if there are any parallel or perpendicular sides.

The top and _____ sides are parallel.

The left and _____ sides are parallel.

> Opposite sides are parallel.

Since there are 4 right angles, the sides that form each right angle are perpendicular.

So, there are _____ pairs of perpendicular sides.

Notice that opposite sides are also equal in length.

A quadrilateral with 4 right angles, opposite sides equal in length, and opposite sides parallel is a *rectangle*. A rectangle is a special kind of quadrilateral.

Online Content at connectED.mcgraw-hill.com

Key Concept Quadrilaterals

A **parallelogram** has opposite sides equal in length and parallel. In addition, opposite angles have the same size.

A **rectangle** has opposite sides equal in length and parallel. It has 4 right angles.

A **rhombus** has opposite sides equal in length and parallel. It has 4 equal sides.

A **square** has opposite sides equal in length and parallel. It has 4 right angles and 4 equal sides.

A **trapezoid** has exactly one pair of parallel sides.

A rectangle, a square, and a rhombus each have all of the attributes, or characteristics, of a parallelogram. So, they are also parallelograms.

Example 2

Classify the quadrilateral in as many ways as possible.
The quadrilateral has opposite sides equal in length and opposite sides parallel. It also has _____ equal sides.

So, it is a _____ and a _____.

Talk MATH

Explain why a square is also a parallelogram.

Guided Practice

1. Classify the quadrilateral in as many ways as possible.

   It is a _____,
   a _____, a _____,
   and a _____.

926 Chapter 14 Geometry

Name _____

Independent Practice

Classify each quadrilateral in as many ways as possible.

2.

3.

4.

_____   _____   _____

_____   _____   _____

Write the type of quadrilateral that best describes each shape. Choose the most specific term.

5.

6.

7.

_____   _____   _____

8. Follow the steps to describe the quadrilateral.

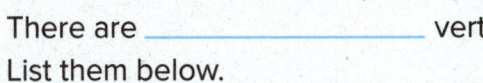

The quadrilateral is a _____ because it has exactly one pair of _____ sides.

There are _____ vertices. List them below.

Vertices: _____

There are _____ line segments. List them below.

Line Segments: _____

Explain why the figure is not a parallelogram.

_____

Lesson 9 Quadrilaterals **927**

Problem Solving

9. A quadrilateral has opposite sides parallel and 4 right angles. Two sides are longer than the others. What is the quadrilateral?

   _____

10. Draw four line segments that form a square.

11. **Processes &Practices 2** **Reason** How are a square and a rhombus alike? How are they different?

    _____

    _____

    _____

Brain Builders

Processes &Practices 7 **Identify Structure** Tell whether each statement is true or false. If the statement is false, draw an example to show why it is false. Then change the statement so that it is true.

12. A rhombus is a square. _____

    _____

13. A rectangle is a parallelogram. _____

    _____

14. **Building on the Essential Question** How can I classify quadrilaterals?

    _____

    _____

    _____

Name ....

MY Homework

Lesson 9

Quadrilaterals

Homework Helper

Need help? connectED.mcgraw-hill.com

Classify each quadrilateral in as many ways as possible.

Quadrilateral 1   **Quadrilateral 2**

Quadrilateral 1 is a **parallelogram** because its opposite sides are equal in length and parallel.

It is also a **rhombus** because it has 4 sides equal in length.

Quadrilateral 2 is a **parallelogram** because its opposite sides are equal in length and parallel.

It is also a **rectangle** because it has 4 right angles.

Practice

Classify each quadrilateral in as many ways as possible.

1.

2.

Lesson 9 My Homework 929

Draw and classify a quadrilateral that fits each description.

3. 4 right angles, opposite sides equal in length and parallel

4. opposite sides equal in length and parallel

Brain Builders

5. **Processes & Practices** **Be Precise** Simon and Chaz each draw a quadrilateral with two sides that are 2 inches long and two sides that are 3 inches long. Their figures are not the same type of quadrilateral. What could the figures be?

Vocabulary Check

6. Explain the difference between a parallelogram and a rhombus.

7. Explain the difference between a square and a rectangle.

8. How is a trapezoid different from the other four types of quadrilaterals you have learned about?

9. **Test Practice** Which quadrilateral does *not* have opposite sides equal in length?

   Ⓐ rhombus   Ⓑ trapezoid   Ⓒ parallelogram   Ⓓ square

Need more practice? Download Extra Practice at connectED.mcgraw-hill.com

Name ....

Lesson 10
Draw Lines of Symmetry

ESSENTIAL QUESTION
How are different ideas about geometry connected?

A figure has **line symmetry** if it can be folded over a line so that one half of the figure matches the other half. This fold line is called the **line of symmetry**.

 Math in My World

Example 1

Determine whether the sign at the right has line symmetry. If it does, draw the line(s) of symmetry on the figure.

 Determine if the figure has line symmetry.
The figure can be folded in half vertically so that the left side matches the right side. It can also be folded in half horizontally so that the top matches the bottom.

So, the figure has line symmetry.

 Draw the line(s) of symmetry.
Draw a vertical line and a horizontal line through the center of the figure.

Some figures have more than one line of symmetry.

The pentagon at the right has five lines of symmetry.

Notice that the arrow ends that are not labeled are the other ends of the arrows that are labeled.

Some figures do not have any lines of symmetry.

The trapezoid at the right does not have any lines of symmetry.

Online Content at ⇒connectED.mcgraw-hill.com

Lesson 10 931

Example 2

Circle the figures that have line symmetry. On those figures, draw all the lines of symmetry.

 Determine which figures have line symmetry. List them.

_____

2 Draw the lines of symmetry on those figures.

Guided Practice

Determine whether each figure has line symmetry. If it does, draw the line(s) of symmetry.

1.

2.

3.

Name one of the capital letters of the alphabet that does not have any line symmetry.

932 Chapter 14 Geometry

Name _____

Independent Practice

Determine whether each figure has line symmetry. Write *yes* or *no*. Draw the line(s) of symmetry on the figures that have line symmetry.

4. _____

5. _____

6. _____

7. _____

Circle the figures that have line symmetry. Cross out the figures that do *not* have line symmetry.

8.

9.

10.

11.

12.

13.

Determine whether the dotted line is a line of symmetry for each figure. Write *yes* or *no*.

14.

15.

16.

_____   _____

Lesson 10 Draw Lines of Symmetry 933

Problem Solving

17. Processes & Practices 2 **Stop and Reflect** Layla saw this sign on her way home from school. Does this sign have line symmetry? Explain.

_____

_____

Draw the other half of each symmetrical shape.

18.

19.

Brain Builders

20. Processes & Practices 4 **Model Math** Draw an object that shows line symmetry and an object that does not show line symmetry. Draw the line of symmetry for the object that shows line symmetry.

21. Building on the Essential Question Name a subject other than math in which symmetry is important. Explain.

_____

_____

_____

Name _____

MY Homework

Lesson 10

Draw Lines of Symmetry

Homework Helper

Need help? connectED.mcgraw-hill.com

Determine whether the hospital sign has line symmetry. If it does, draw the line(s) of symmetry on the figure.

 Determine if the figure has line symmetry.
The figure can be folded in half vertically so that the left side matches the right side.

It can also be folded in half horizontally so that the top matches the bottom.

So, the figure has line symmetry.

 Draw the lines of symmetry.
Draw a vertical line through the center of the figure.

Draw a horizontal line through the center of the figure.

There are 2 lines of symmetry.

Practice

Determine whether each figure has line symmetry. Write *yes* or *no*. Draw the line(s) of symmetry on the figures that have line symmetry.

1. _____

2. _____

3. _____

4. _____

Lesson 10 My Homework 935

Determine whether the dotted line is a line of symmetry for each figure. Write *yes* or *no*.

5. _____

6. _____

Draw the other half of each symmetrical shape.

7.

8.

Brain Builders

9. **Processes & Practices 4** **Model Math** Vince wrote all the capital letters of the alphabet. Which of the letters have more than one line of symmetry?

_____

Vocabulary Check

Choose the correct word(s) to complete each sentence.

      line of symmetry   line symmetry

10. If a figure can be folded into identical halves, it has

_____.

11. The fold is the _____.

12. **Test Practice** Does the sign have more or less lines of symmetry than a rectangle?

   Ⓐ 2 more   Ⓒ 1 less

   Ⓑ 1 more   Ⓓ 2 less

Name _____

Lesson 11
Problem-Solving Investigation
STRATEGY: Make a Model

ESSENTIAL QUESTION
How are different ideas about geometry connected?

Learn the Strategy

Maya has a square piece of paper. She folds it in half so there are two triangle-shaped parts. She folds it in half again so there are four triangle-shaped parts. When she unfolds the paper, how many right angles are shown?

1 Understand

What facts do you know?

Maya folds a piece of paper in half _____ times diagonally.

What do you need to find?

the number of right angles when the paper is unfolded

2 Plan

I will make a model to find the answer.

3 Solve

Use a square piece of paper. Follow the same steps that Maya followed.

Fold the paper.

Fold it again.

Open it up.

Count the number of right angles.

So, there are _____ right angles.

4 Check

Does your answer make sense? Explain.

_____

Online Content at **connectED.mcgraw-hill.com**

Lesson 11   937

Practice the Strategy

Sam has a card that is a quadrilateral. All sides are equal length. One of the angles measures 60°. Which quadrilateral is Sam's card?

1 Understand

What facts do you know?

_____

_____

What do you need to find?

_____

_____

2 Plan

_____

_____

3 Solve

4 Check

Does your answer make sense? Explain.

_____

_____

Name

Apply the Strategy

Solve each problem by making a model.

1. **Processes & Practices 7** **Identify Structure** Mary Anne is making a pattern with quadrilaterals. She put squares in the first row, parallelograms in the second row, and trapezoids in the third row. She repeats this pattern four times. Which quadrilateral does she use in the tenth row?

Brain Builders

2. **Processes & Practices 4** **Model Math** Draw two lines on the square below so that three right triangles are formed. How many right angles are formed?

3. **Processes & Practices 6** **Explain to a Friend** The first time a pizza is cut in half, there are 2 pieces. The second time each piece is cut in half, there are 4 pieces. The third time each piece is cut in half, there are 8 pieces. How many pieces will there be after each piece is cut the fifth time? Explain how you solved the problem.

Lesson 11 Problem-Solving Investigation 939

Review the Strategies

Use any strategy to solve each problem.
- Make a model.
- Find a pattern.
- Make a table.
- Guess, check, and revise.

4. Processes & Practices 8 **Look for a Pattern** Mandy does one chore a day for her allowance. One day she washes dishes, the next she walks the dog, and the next she folds laundry. If she starts this cycle of chores on a Monday, which chore will she be doing the following Monday?

5. Kendra took photographs at the park. She photographed 20 dogs and owners in all. If there was a total of 64 legs, how many dogs and owners were there?

6. Processes & Practices 7 **Identify Structure** Corey drew 8 shapes. Four are squares, two are triangles, and the rest are parallelograms. Write two fractions that name the part of the shapes that are quadrilaterals.

7. Processes & Practices 1 **Make Sense of Problems** Mr. Lawrence used 100 tiles when he put a new floor in the kitchen. Thirty-four tiles are squares and 16 tiles are rectangles. Write the total amount of tiles that are squares and rectangles as a decimal and a fraction.

940 Chapter 14 Geometry

Name _____

MY Homework

Lesson 11

Problem Solving: Make a Model

Homework Helper

Need help? connectED.mcgraw-hill.com

Bernie has three identical blocks. Each block has the shape of an acute triangle. If Bernie puts two of the blocks together, what quadrilateral does he form? If he joins the third block to the first two, what new quadrilateral does he form?

1 Understand

What facts do you know?

Bernie has three identical blocks shaped like acute triangles.

What do you need to find?

the quadrilaterals that Bernie will form by joining the blocks

2 Plan

Make a model to find the answer.

3 Solve

Draw an acute triangle. Join an identical acute triangle to the first one to form a quadrilateral.

The quadrilateral is a parallelogram.

The new quadrilateral is a trapezoid.

Draw a third acute triangle joined to the first two triangles to form a new quadrilateral.

So, Bernie forms a parallelogram and a trapezoid.

4 Check

The figures match the information in the problem. So, the answer makes sense.

Lesson 11 My Homework 941

Problem Solving

Solve each problem by making a model.

1. **Processes & Practices 7** **Identify Structure** A quadrilateral has sides that measure 12 centimeters, 10 centimeters, 10 centimeters, and 5 centimeters. The side that measures 12 centimeters is parallel to the side that measures 5 centimeters. What is the figure?

2. **Processes & Practices 6** **Explain to a Friend** Katerina draws a square. She wants to draw one line to divide the square into 2 triangles. Is it possible for her to divide the square into 2 obtuse triangles? Explain.

Brain Builders

3. Is it possible for a quadrilateral to have only 2 right angles? If so, draw an example and classify the other two angles as acute, right, or obtuse.

4. **Processes & Practices 7** **Identify Structure** Kirby drew a quadrilateral with four right angles, a side that measures 9 centimeters, and a perimeter that measures 30 centimeters. What is the figure? What is the measure of each side?

5. **Processes & Practices 6** **Be Precise** Is it possible for a trapezoid to have only 3 right angles? Explain. Include a sketch to support your answer.

Review

Chapter 14
Geometry

Vocabulary Check

Draw an example of each vocabulary word.

1. acute angle	2. acute triangle	3. intersecting lines
4. line	5. line segment	6. obtuse angle
7. obtuse triangle	8. parallel lines	9. perpendicular lines
10. ray	11. right triangle	12. line symmetry

Concept Check

Describe each figure. Use *parallel*, *perpendicular*, or *intersecting*. Use the most specific term.

13.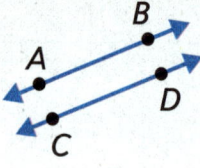

    $\overleftrightarrow{AB}$ and $\overleftrightarrow{CD}$ are

    _____.

14.

    $\overleftrightarrow{ST}$ and $\overleftrightarrow{UV}$ are

    _____.

15. Draw an angle with a measure greater than $\frac{1}{4}$ turn.

16. Classify the angle as *right*, *acute*, or *obtuse*.

    _____

17. Classify the triangle as *acute*, *right*, or *obtuse*.

    _____

18. Write the type of quadrilateral that best describes the shape at the right.

    _____

Draw the line(s) of symmetry in each figure.

19.

20.

944 Chapter 14 Geometry

Problem Solving

21. Brent started his homework at 4:00 p.m. He completed it at the time shown. Write how far the minute hand has turned in degrees and as a fraction of a full turn.

22. Katie drew the first letter of her name. What types of angles did she draw? Draw and label a diagram to support your answer.

Brain Builders

23. Sergio saw an angle that measures greater than 0° and less than 90°. What type of angle did he see?

24. **Test Practice** Levi's school is at the corner of High Street and Second Avenue. The corner forms a right angle. How might he describe the way the streets meet at his school?

   Ⓐ The streets are parallel.
   Ⓑ The streets are intersecting, but not perpendicular.
   Ⓒ The streets never meet.
   Ⓓ The streets are perpendicular.

Reflect

Chapter 14
Answering the
ESSENTIAL QUESTION

Use what you learned about geometry to complete the graphic organizer.

ESSENTIAL QUESTION

How are different ideas about geometry connected?

- line segments
- angles
- symmetry

Draw a geometric figure. Use vocabulary words to label the figure.

Now reflect on the **ESSENTIAL QUESTION** Write your answer below.

Name _____ Date _____

Score _____

Performance Task

Engineer Ellen

As an engineer working for the city, Ellen has to know everything about lines, angles, and shapes. She has to use the correct structures to help build bridges and roads, so the blueprints must be precise.

Show all your work to receive full credit.

Part A

Part of a blueprint of a bridge shows this support:

On the blueprint, it says that angle X is supposed to be exactly 52°. Is this correct? If not, draw the correct angle.

Part B

Ellen has to draw a new blueprint that includes a triangle with angles of 60°, 55°, and 65°. Draw the triangle and label each angle. Then write the name of the triangle.

Online Content at connectED.mcgraw-hill.com

Performance Task   946PT1

Part C

As part of her duties, Ellen has to keep maps of the city up to date. This is one of those maps. Identify a set of parallel lines, a set of perpendicular lines, and a trapezoid on the map.

Part D

Ellen has to know about lines of symmetry so she can be sure parts of the bridge will balance. Which of these figures have a line of symmetry? Which have more than 1 line of symmetry? Draw all the lines of symmetry you can find.

946PT2 **Chapter 14** Geometry

Glossary/Glosario

Go to the eGlossary to find out more about these words in the following 13 languages:

Arabic • Bengali • Brazilian Portuguese • Cantonese • English • Haitian Creole
Hmong • Korean • Russian • Spanish • Tagalog • Urdu • Vietnamese

English	Spanish/Español
acute angle An *angle* with a measure greater than 0° and less than 90°.	**ángulo agudo** *Ángulo* que mide más de 0° y menos de 90°.
acute triangle A *triangle* with all three *angles* less than 90°.	**triángulo acutángulo** *Triángulo* cuyos tres *ángulos* miden menos de 90°.
add (adding, addition) An *operation* on two or more *addends* that results in a *sum*. $9 + 3 = 12$	**suma (sumar)** *Operación* de dos o más *sumandos* que da como resultado una *suma*. $9 + 3 = 12$
addend Any numbers being *added* together.	**sumando** Cualquier número que se *suma* a otro.
algebra A branch of mathematics that uses symbols, usually letters, to explore relationships between quantities.	**álgebra** Rama de las matemáticas en la que se usan símbolos, generalmente letras, para explorar relaciones entre cantidades.

Online Content at connectED.mcgraw-hill.com

Aa

angle A figure that is formed by two *rays* with the same *endpoint*.

ángulo Figura formada por dos *semirrectas* con el mismo *extremo*.

area The number of *square units* needed to cover the inside of a region or plane figure without any overlap.

area = 6 square units

área Cantidad de *unidades cuadradas* necesarias para cubrir el interior de una región o figura plana sin superposiciones.

área = 6 unidades cuadradas

Associative Property of Addition The property that states that the grouping of the *addends* does not change the *sum*.

$(4 + 5) + 2 = 4 + (5 + 2)$

propiedad asociativa de la suma Propiedad que establece que la agrupación de los *sumandos* no altera la *suma*.

$(4 + 5) + 2 = 4 + (5 + 2)$

Associative Property of Multiplication The property that states that the grouping of the *factors* does not change the *product*.

$3 \times (6 \times 2) = (3 \times 6) \times 2$

propiedad asociativa de la multiplicación Propiedad que establece que la agrupación de los *factores* no altera el *producto*.

$3 \times (6 \times 2) = (3 \times 6) \times 2$

Bb

bar graph A graph that compares *data* by using bars of different *lengths* or heights to show the values.

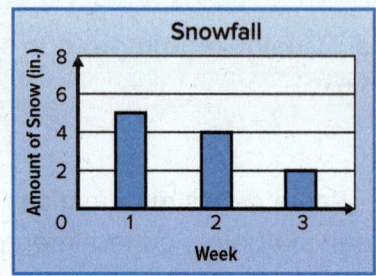

gráfica de barras Gráfica en la que se comparan los *datos* usando barras de distintas *longitudes* o alturas para mostrar los valores.

capacity The amount of liquid a container can hold.

centimeter (cm) A *metric* unit for measuring *length*.

$$100 \text{ centimeters} = 1 \text{ meter}$$

circle A closed figure in which all points are the same distance from a fixed point, called the center.

Commutative Property of Addition The property that states that the order in which two numbers are *added* does not change the *sum*.

$$12 + 15 = 15 + 12$$

Commutative Property of Multiplication The property that states that the order in which two numbers are *multiplied* does not change the *product*.

$$7 \times 2 = 2 \times 7$$

compatible numbers Numbers in a problem or related numbers that are easy to work with mentally.

720 and 90 are compatible numbers for *division* because $72 \div 9 = 8$.

composite number A whole number that has more than two factors.

12 has the factors 1, 2, 3, 4, 6, and 12.

---

capacidad Cantidad que puede contener un recipiente, medida en unidades de volumen.

centímetro (cm) Unidad métrica de *longitud*.

$$100 \text{ centímetros} = 1 \text{ metro}$$

círculo Figura cerrada en la cual todos los puntos equidistan de un punto fijo llamado centro.

propiedad conmutativa de la suma Propiedad que establece que el orden en el que se *suman* dos o más números no altera la *suma*.

$$12 + 15 = 15 + 12$$

propiedad conmutativa de la multiplicación Propiedad que establece que el orden en el que se *multiplican* dos o más números no altera el *producto*.

$$7 \times 2 = 2 \times 7$$

números compatibles Números en un problema o números relacionados con los cuales es fácil trabajar mentalmente.

$720 \div 90$ es una división que usa números compatibles porque $72 \div 9 = 8$.

número compuesto Número natural con más de dos factores.

12 tiene los factores 1, 2, 3, 4, 6 y 12.

Cc

congruent figures Two figures having the same size and the same shape.

convert To change one unit to another.

cup (c) A *customary* unit of *capacity* equal to 8 fluid ounces.

customary system The measurement system most often used in the United States. Units include *foot*, *pound*, and *quart*.

figuras congruentes Dos figuras con la misma forma y el mismo tamaño.

convertir Cambiar de una unidad a otra.

taza (tz) Unidad *usual* de *capacidad* que equivale a 8 onzas líquidas.

sistema usual Conjunto de unidades de medida de uso más frecuente en Estados Unidos. Incluye unidades como *el pie*, *la libra* y *el cuarto*.

Dd

data Numbers or symbols, sometimes collected from a *survey* or experiment, to show information. Datum is singular; data is plural.

decimal A number with one or more digits to the right of the decimal point, such as 8.37 or 0.05.

decimal equivalents Decimals that represent the same number.

              0.3 and 0.30

decimal point A period separating the ones and the *tenths* in a decimal number.

              0.8 or $3.77

decompose To break a number into different parts.

degree (°) **a.** A unit for measuring angles. **b.** A unit of measure used to describe temperature.

datos Números o símbolos que muestran información, algunas veces recopilados a partir de una *encuesta* o un experimento.

decimal Número con uno o más dígitos a la derecha del *punto decimal*, como 8.37 o 0.05.

decimales equivalentes Decimales que representan el mismo número.

              0.3 y 0.30

punto decimal Punto que separa las unidades de las *décimas* en un número decimal.

              0.8 o $3.77

descomponer Separar un número en diferentes partes.

grado (°) **a.** Unidad que se usa para medir ángulos. **b.** Unidad de medida que se usa para describir la temperatura.

Glossary/Glosario

Dd

denominator The bottom number in a *fraction*.

In $\frac{5}{6}$, 6 is the denominator.

digit A symbol used to write numbers. The ten digits are 0, 1, 2, 3, 4, 5, 6, 7, 8, and 9.

Distributive Property To *multiply* a *sum* by a number, multiply each *addend* by the number and *add* the *products*.

$$4 \times (1 + 3) = (4 \times 1) + (4 \times 3)$$

dividend A number that is being *divided*.

$3\overline{)19}$ 19 is the dividend.

division (divide) An *operation* on two numbers in which the first number is split into the same number of equal groups as the second number.

divisor The number by which the *dividend* is being *divided*.

$3\overline{)19}$ 3 is the divisor.

denominador El número de abajo en una *fracción*.

En $\frac{5}{6}$, 6 es el denominador.

dígito Símbolo que se usa para escribir los números. Los diez dígitos son 0, 1, 2, 3, 4, 5, 6, 7, 8 y 9.

propiedad distributiva Para *multiplicar* una *suma* por un número, multiplica cada *sumando* por ese número y luego *suma* los *productos*.

$$4 \times (1 + 3) = (4 \times 1) + (4 \times 3)$$

dividendo Número que se *divide*.

$3\overline{)19}$ 19 es el dividendo.

división (dividir) *Operación* entre dos números en la que el primer número se separa en tantos grupos iguales como indica el segundo número.

divisor Número entre el cual se *divide* el *dividendo*.

$3\overline{)19}$ 3 es el divisor.

Ee

elapsed time The amount of time that has passed from beginning to end.

endpoint The point at either end of a *line segment* or the point at the beginning of a *ray*.

tiempo transcurrido Cantidad de tiempo que ha pasado entre el principio y el fin de algo.

extremo Punto que se encuentra en cualquiera de los dos lados en que termina un *segmento de recta* o al principio de una *semirrecta*.

Ee

equation A sentence that contains an equals sign (=), showing that two *expressions* are equal.

equilateral triangle A *triangle* with three *congruent* sides.

equivalent fractions *Fractions* that represent the same number.

$$\frac{3}{4} = \frac{6}{8}$$

estimate A number close to an exact value. An estimate indicates *about* how much.

47 + 22 is about
50 + 20 or 70.

expanded form/expanded notation The representation of a number as a *sum* that shows the value of each digit.

536 is written as 500 + 30 + 6.

expression A combination of numbers, *variables,* and at least one *operation*.

ecuación Espresión matemática que contiene el signo igual, (=), que indica que las dos *expresiones* son iguales.

triángulo equilátero *Triángulo* con tres lados *congruentes*.

fracciones equivalentes *Fracciones* que representan el mismo número.

$$\frac{3}{4} = \frac{6}{8}$$

estimación Número cercano a un valor exacto. Una estimación indica una cantidad approximada.

47 + 22 es aproximadamente
50 + 20 o 70.

forma desarrollada/notación desarrollada Representación de un número como la *suma* del valor de cada dígito.

536 puede escribirse como 500 + 30 + 6.

expresión Combinación de números, *variables* y por lo menos una *operación*.

fact family A group of related facts using the same numbers.

5 + 3 = 8	5 × 3 = 15
3 + 5 = 8	3 × 5 = 15
8 − 3 = 5	15 ÷ 3 = 5
8 − 5 = 3	15 ÷ 5 = 3

familia de operaciones Grupo de operaciones relacionadas que usan los mismos números.

5 + 3 = 8	5 × 3 = 15
3 + 5 = 8	3 × 5 = 15
8 − 3 = 5	15 ÷ 3 = 5
8 − 5 = 3	15 ÷ 5 = 3

Ff

factor A number that *divides* a whole number evenly. Also a number that is *multiplied* by another number.

factor pairs The two factors that are multiplied to find a product.

fluid ounce (fl oz) A *customary* unit of *capacity*.

foot (ft) A *customary* unit for measuring *length*. Plural is feet.

$$1 \text{ foot} = 12 \text{ inches}$$

formula An *equation* that shows the relationship between two or more quantities.

fraction A number that represents part of a whole or part of a set.

$$\frac{1}{2}, \frac{1}{3}, \frac{1}{4}, \frac{3}{4}$$

frequency table A table for organizing a set of *data* that shows the number of times each result has occurred.

factor Número entre el que se *divide* otro número natural sin dejar residuo. También es cada uno de los números en una multiplicación.

pares de factores Los dos factores que se multiplican para hallar un producto.

onza líquida (oz líq) Unidad *usual* de *capacidad*.

pie (pie) Unidad *usual* de *longitud*.

$$1 \text{ pie} = 12 \text{ pulgadas}$$

fórmula *Ecuación* que muestra la relación entre dos o más cantidades.

fracción Número que representa una parte de un todo o una parte de un conjunto.

$$\frac{1}{2}, \frac{1}{3}, \frac{1}{4}, \frac{3}{4}$$

tabla de frecuencias Tabla para organizar un conjunto de *datos* que muestra el número de veces que se ha obtenido cada resultado.

gallon (gal) A *customary* unit for measuring *capacity* for liquids.

$$1 \text{ gallon} = 4 \text{ quarts}$$

gram (g) A *metric* unit for measuring *mass*.

galón (gal) Unidad *usual* de *capacidad* de líquidos.

$$1 \text{ galón} = 4 \text{ cuartos}$$

gramo (g) Unidad *métrica* de *masa*.

Gg

Greatest Common Factor (GCF) The greatest of the common *factors* of two or more numbers.

The greatest common factor of 12, 18, and 30 is 6.

máximo común divisor (M.C.D.) El mayor de los *factores* comunes de dos o más números.

El máximo común divisor de 12, 18 y 30 es 6.

Hh

hexagon A *polygon* with six sides and six *angles*.

hexágono *Polígono* con seis lados y seis *ángulos*.

hundredth A *place-value* position. One of one hundred equal parts.

In the number 0.05, 5 is in the hundredths place.

centésima *Valor posicional.* Una de cien partes iguales.

En el número 4.57, 7 está en el lugar de las centésimas.

Ii

Identity Property of Addition For any number, zero plus that number is the number.

$3 + 0 = 3$ or $0 + 3 = 3$

propiedad de identidad de la suma Para todo número, cero más ese número da como resultado ese mismo número.

$3 + 0 = 3$ o $0 + 3 = 3$

Identity Property of Multiplication If you *multiply* a number by 1, the *product* is the same as the given number.

$8 \times 1 = 8 = 1 \times 8$

propiedad de identidad de la multiplicación Si *multiplicas* un número por 1, el *producto* es igual al número dado.

$8 \times 1 = 8 = 1 \times 8$

improper fraction A *fraction* with a *numerator* that is greater than or equal to the *denominator*.

$\frac{17}{3}$ or $\frac{5}{5}$

fracción impropia *Fracción* con un *numerador* igual al *denominador* o mayor que él.

$\frac{17}{3}$ o $\frac{5}{5}$

input A quantity that is changed to produce an output.

intersecting lines *Lines* that meet or cross at a point.

is equal to (=) Having the same value. The (=) sign is used to show two numbers or *expressions* are equal.

is greater than (>) An inequality relationship showing that the number on the left of the symbol is greater than the number on the right.

$5 > 3$ 5 is greater than 3.

is less than (<) An inequality relationship showing that the number on the left side of the symbol is less than the number on the right side.

$4 < 7$ 4 is less than 7.

isosceles triangle A *triangle* with at least 2 sides of the same *length*.

entrada Cantidad que se modifica y produce un valor de salida.

rectas secantes *Rectas* que se intersecan o se cruzan en un punto común.

es igual a (=) Que tiene el mismo valor. Con el signo = se muestra que dos números o *expresiones* son iguales.

es mayor que (>) Relación de desigualdad que muestra que el número a la izquierda del signo es más grande que el número a la derecha.

$5 > 3$ 5 es mayor que 3.

es menor que (<) Relación de desigualdad que muestra que el número a la izquierda del signo es más pequeño que el número a la derecha.

$4 < 7$ 4 es menor que 7.

triángulo isósceles *Triángulo* que tiene por lo menos 2 lados del mismo *largo*.

Kk

kilogram (kg) A *metric* unit for measuring *mass*.

kilometer (km) A *metric* unit for measuring *length*.

kilogramo (kg) Unidad *métrica* de *masa*.

kilómetro (km) Unidad *métrica* de *longitud*.

Ll

length The measurement of a *line* between two points.

longitud Medida de la distancia entre dos puntos.

like fractions *Fractions* that have the same *denominator*.

$\frac{1}{5}$ and $\frac{2}{5}$

fracciones semejantes *Fracciones* que tienen el mismo *denominador*.

$\frac{1}{5}$ y $\frac{2}{5}$

line A straight set of points that extend in opposite directions without ending.

recta Conjunto de puntos alineados que se extiende sin fin en direcciones opuestas.

line of symmetry A *line* on which a figure can be folded so that its two halves match exactly.

eje de simetría *Recta* sobre la cual se puede doblar una figura de manera que sus mitades coincidan exactamente.

line plot A graph that uses columns of Xs above a *number line* to show frequency of *data*.

diagrama lineal Gráfica que tiene columnas de X sobre una *recta numérica* para representar la frecuencia de los *datos*.

line segment A part of a *line* between two *endpoints*. The *length* of the line segment can be measured.

line symmetry A figure has *line symmetry* if it can be folded so that the two parts of the figure match, or are *congruent*.

liter (L) A *metric* unit for measuring *volume* or *capacity*.

1 liter = 1,000 milliliters

segmento de recta Parte de una *recta* entre dos *extremos*. La *longitud* de un segmento de recta se puede medir.

simetría axial Una figura tiene *simetría axial* si puede doblarse de modo que las dos partes de la figura coincidan de manera exacta.

litro (L) Unidad *métrica* de *volumen* o *capacidad*.

1 litro = 1,000 mililitros

mass The amount of matter in an object. Two examples of units of measure would be gram and kilogram.

meter (m) A *metric* unit for measuring *length*.

metric system (SI) The decimal system of measurement. Includes units such as *meter*, *gram*, and *liter*.

mile (mi) A *customary* unit of measure for *length*.

1 mile = 5,280 feet

milliliter (mL) A *metric* unit for measuring *capacity*.

1,000 milliliters = 1 liter

millimeter (mm) A *metric* unit for measuring *length*.

1,000 millimeters = 1 meter

masa Cantidad de materia en un cuerpo. El gramo y el kilogramo son dos ejemplos de unidades que se usan para medir la masa.

metro (m) Unidad *métrica* de *longitud*.

sistema métrico (SI) Sistema decimal de medidas que se basa en potencias de 10 y que incluye unidades como *el metro*, *el gramo* y *el litro*.

milla (mi) Unidad *usual* de *longitud*.

1 milla = 5,280 pies

mililitro (mL) Unidad *métrica* de *capacidad*.

1,000 mililitros = 1 litro

milímetro (mm) Unidad *métrica* de *longitud*.

1,000 milímetros = 1 metro

Online Content at connectED.mcgraw-hill.com

Mm

minuend The first number in a *subtraction* sentence from which a second number is to be subtracted.

mixed number A number that has a *whole number* part and a *fraction* part.

$$6\frac{3}{4}$$

multiple A multiple of a number is the *product* of that number and any whole number.

15 is a multiple of 5 because $3 \times 5 = 15$.

multiply (multiplication) An *operation* on two numbers to find their *product*. It can be thought of as repeated *addition*.

minuendo El primer número en un enunciado de *resta* del cual se restará un segundo número

8 − 3 = 5
minuendo sustraendo diferencia

número mixto Número formado por un *número natural* y una parte *fraccionaria*.

$$6\frac{3}{4}$$

múltiplo Un múltiplo de un número es el *producto* de ese número y cualquier otro número natural.

15 es múltiplo de 5 porque $3 \times 5 = 15$.

multiplicar (multiplicación) *Operación* entre dos números para hallar su *producto*. También se puede interpretar como una *suma* repetida.

Nn

nonnumeric pattern Patterns that do not use numbers.

number line A *line* with numbers on it in order at regular intervals.

numerator The number above the bar in a *fraction*; the part of the fraction that tells how many of the equal parts are being used.

numeric pattern Patterns that use numbers.

patrón no numérico Patrón que no usa números.

recta numérica *Recta* con números ordenados a intervalos regulares.

numerador El número que está encima de la barra de *fracción*; la parte de la fracción que indica cuántas de las partes iguales en que se divide el entero se están usando.

patrón numérico Patrón que usa números.

Oo

obtuse angle An *angle* that measures greater than 90° but less than 180°.

obtuse triangle A *triangle* with one *obtuse angle*.

octagon A *polygon* with 8 sides and 8 *angles*.

operation A mathematical process such as *addition* (+), *subtraction* (−), *multiplication* (×), or *division* (÷).

order of operations Rules that tell what order to follow when evaluating an *expression*:
(1) Do the *operations* in *parentheses* first.
(2) *Multiply* and *divide* in order from left to right.
(3) *Add* and *subtract* in order from left to right.

ounce (oz) A *customary* unit to measure *weight* or *capacity*.

output The result of an input quantity being changed.

ángulo obtuso *Ángulo* que mide más de 90° pero menos de 180°.

triángulo obtusángulo *Triángulo* con un *ángulo obtuso*.

octágono *Polígono* de 8 lados y 8 *ángulos*.

operación Proceso matemático como la *suma* (+), la *resta* (−), la *multiplicación* (×) o la *división* (÷).

orden de las operaciones Reglas que te indican qué orden seguir cuando evalúas una *expresión*:
(1) Resuelve primero las *operaciones* dentro de los *paréntesis*.
(2) *Multiplica* o *divide* en orden de izquierda a derecha.
(3) *Suma* o *resta* en orden de izquierda a derecha.

onza (oz) Unidad *usual* de *peso* o *capacidad*.

salida Resultado que se obtiene al modificar un valor de entrada.

Pp

parallel lines *Lines* that are the same distance apart. Parallel lines do not meet.

rectas paralelas *Rectas* separadas por la misma distancia en cualquier punto. Las rectas paralelas no se intersecan.

parallelogram A *quadrilateral* with four sides in which each pair of opposite sides are *parallel* and equal in *length*.

paralelogramo *Cuadrilátero* en el que cada par de lados opuestos son *paralelos* y tienen la misma *longitud*.

parentheses The enclosing symbols ( ), which indicate that the terms within are a unit.

paréntesis Los signos ( ) con que se encierran los términos, para indicar que cuando están adentro, forman una unidad.

partial products A multiplication method in which the products of each place value are found separately, and then added together.

productos parciales Método de multiplicación por el cual los productos de cada valor posicional se hallan por separado y luego se suman entre sí.

partial quotients A dividing method in which the dividend is separated into sections that are easy to divide.

cocientes parciales Método de división por el cual el dividendo se separa en secciones que son fáciles de dividir.

pattern A sequence of numbers, figures, or symbols that follows a rule or design.

2, 4, 6, 8, 10

patrón Sucesión de números, figuras o símbolos que sigue una regla o un diseño.

2, 4, 6, 8, 10

pentagon A *polygon* with five *sides* and five *angles*.

pentágono *Polígono* de cinco *lados* y cinco *ángulos*.

GL14 Glossary/Glosario

Pp

percent A ratio that compares a number to 100.

perimeter The distance around a shape or region.

period The name given to each group of three digits on a place-value chart.

perpendicular lines *Lines* that meet or cross each other to form *right angles*.

pint (pt) A *customary* unit for measuring *capacity*.

1 pint = 2 cups

place value The value given to a *digit* by its position in a number.

point An exact location in space that is represented by a dot.

polygon A closed *plane figure* formed using *line segments* that meet only at their *endpoints*.

pound (lb) A *customary* unit to measure *weight* or *mass*.

1 pound = 16 ounces

porcentaje Razón que compara un número con el 100.

perímetro Distancia alrededor de una figura o región.

período Nombre dado a cada grupo de tres dígitos en una tabla de valor posicional.

rectas perpendiculares *Rectas* que se intersecan o cruzan formando *ángulos rectos*.

pinta (pt) Unidad *usual* de *capacidad*.

1 pinta = 2 tazas

valor posicional Valor dado a un *dígito* según su posición en un número.

punto Ubicación exacta en el espacio que se representa con una marca puntual.

polígono *Figura plana* cerrada formada por *segmentos de recta* que solo se unen en sus *extremos*.

libra (lb) Unidad *usual* de *peso* o *masa*.

1 libra = 16 onzas

Pp

prime number A whole number with exactly two *factors*, 1 and itself.

7, 13, and 19

product The answer or result of a *multiplication* problem. It also refers to expressing a number as the *product* of its *factors*.

protractor An instrument used to measure angles.

número primo Número natural que tiene exactamente dos *factores*: 1 y sí mismo.

7, 13 y 19

producto Respuesta o resultado de un problema de *multiplicación*. Además, un número puede expresarse como el *producto* de sus *factores*.

transportador Instrumento con el que se miden los ángulos.

Qq

quadrilateral A shape that has 4 sides and 4 *angles*.

square, rectangle, and parallelogram

cuadrilátero Figura que tiene 4 lados y 4 *ángulos*.

cuadrado, rectángulo y paralelogramo

quart (qt) A *customary* unit for measuring *capacity*.

1 quart = 4 cups

quotient The result of a *division* problem.

cuarto (ct) Unidad *usual* de *capacidad*.

1 cuarto = 4 tazas

cociente Respuesta o resultado de un problema de *división*.

Rr

ray A part of a *line* that has one *endpoint* and extends in one direction without ending.

semirrecta Parte de una *recta* que tiene un *extremo* y que se extiende sin fin en una dirección.

rectangle A *quadrilateral* with four *right angles*; opposite sides are equal and *parallel*.

regroup To use place value to exchange equal amounts when renaming a number.

remainder The number that is left after one whole number is *divided* by another.

repeated subtraction To subtract the same number over and over until you reach 0.

rhombus A *parallelogram* with four *congruent* sides.

right angle An *angle* with a measure of 90°.

right triangle A *triangle* with one *right angle*.

rectángulo *Cuadrilátero* con cuatro *ángulo rectos*; los lados opuestos son iguales y *paralelos*.

reagrupar Usar el valor posicional para expresar una cantidad de otra manera.

residuo Número que queda después de *dividir* un número natural entre otro.

resta repetida Procedimiento por el que se resta un número una y otra vez hasta llegar a 0.

rombo *Paralelogramo* con cuatro lados *congruentes*.

ángulo recto *Ángulo* que mide 90°.

triángulo rectángulo *Triángulo* con un *ángulo recto*.

Rr

round To change the value of a number to one that is easier to work with. To find the nearest value of a number based on a given *place value*.

redondear Cambiar el valor de un número a uno con el cual es más fácil trabajar. Hallar el valor más cercano a un número basándose en un *valor posicional* dado.

rule A statement that describes a relationship between numbers or objects.

regla Enunciado que describe una relación entre números u objetos.

Ss

second A unit of time.
60 seconds = 1 minute

segundo Unidad de tiempo.
60 segundos = 1 minuto

sequence The ordered arrangement of terms that make up a pattern.

secuencia Disposición ordenada de términos que forman un patrón.

simplest form A *fraction* in which the *numerator* and the *denominator* have no common *factor* greater than 1.
$\frac{3}{5}$ is the simplest form of $\frac{6}{10}$.

mínima expresión *Fracción* en la que el *numerador* y el *denominador* no tienen un *factor* común mayor que 1.
$\frac{3}{5}$ es la mínima expresión de $\frac{6}{10}$.

solve To replace a *variable* with a value that results in a true sentence.

resolver Despejar una *variable* y reemplazarla por un valor que haga que la ecuación sea verdadera.

square A *rectangle* with four *congruent* sides.

cuadrado *Rectángulo* de cuatro lados *congruentes*.

square unit A unit for measuring *area*.

unidad cuadrada Unidad para medir el *área*.

standard form/standard notation The usual way of writing a number that shows only its *digits*, no words.
537   89   1,642

forma estándar/notación estándar Manera habitual de escribir un número usando solo sus *dígitos*, sin usar palabras.
537   89   1,642

Ss

subtract (subtraction) An *operation* on two numbers that tells the *difference*, when some or all are taken away. Subtraction is also used to compare two numbers.

$$14 - 8 = 6$$

subtrahend A number that is *subtracted* from another number.

$$14 - 5 = 9$$
↑
subtrahend

sum The answer to an *addition* problem.

survey A method of collecting *data*.

restar (resta) *Operación* con dos números que indica la *diferencia*, entre ellos. Puede usarse para quitar una cantidad de otra o para comparar dos números.

$$14 - 8 = 6$$

sustraendo Un número que se *resta* de otro número.

$$14 - 5 = 9$$
↑
sustraendo

suma Respuesta o resultado que se obtiene al sumar.

encuesta Método para recopilar *datos*.

tally chart A way to keep track of *data* using *tally marks* to record the number of responses or occurrences.

What is Your Favorite Color?	
Color	Tally
Blue	|||| |||
Green	||||

tally mark(s) A mark made to keep track of and display *data* recorded from a *survey*.

tenth One of ten equal parts, or $\frac{1}{10}$.

term Each number in a numeric pattern.

tabla de conteo Manera de llevar la cuenta de los *datos* usando *marcas de conteo* para anotar el número de respuestas o sucesos.

¿Cuál es tu color favorito?	
Color	Conteo
azul	|||| |||
verde	||||

marca de conteo Marca que se hace para llevar un registro y representar *datos* recopilados en una *encuesta*.

décima Una de diez partes iguales o $\frac{1}{10}$.

término Cada número en un patrón numérico.

Tt

thousandth(s) One of a thousand equal parts, or $\frac{1}{1000}$. Also refers to a *place value* in a *decimal* number. In the *decimal* 0.789, the 9 is in the thousandths place.

milésima Una de mil partes iguales o $\frac{1}{1000}$. También se refiere a un *valor posicional* en un número *decimal*. En el *decimal* 0.789, el 9 está en el lugar de las milésimas.

three-dimensional figure A solid figure has three dimensions: *length, width*, and *height*.

figura tridimensional Figura sólida que tiene tres dimensiones: *largo, ancho* y *alto*.

ton (T) A *customary* unit to measure *weight*.

1 ton = 2,000 pounds

tonelada (T) Unidad *usual* de *peso*.

tonelada = 2,000 libras

trapezoid A *quadrilateral* with exactly one pair of *parallel* sides.

trapecio *Cuadrilátero* con exactamente un par de lados *paralelos*.

triangle A *polygon* with three sides and three *angles*.

triángulo *Polígono* con tres lados y tres *ángulos*.

two-dimensional figure A figure that lies entirely within one plane.

figura bidimensional Figura que puede representarse en un plano.

Uu

unit square A square with a side length of one unit.

cuadrado unitario Cuadrado cuyos lados miden una unidad de longitud.

unknown The amount that has not been identified.

incógnita La cantidad que no ha sido identificada.

variable A letter or symbol used to represent an unknown quantity.

Venn diagram A diagram that uses *circles* to display elements of different sets. Overlapping *circles* show common elements.

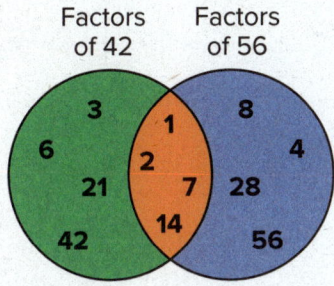

vertex The point where two *rays* meet in an *angle*.

variable Letra o símbolo que se usa para representar una cantidad desconocida.

diagrama de Venn Diagrama con *círculos* para mostrar elementos de diferentes conjuntos. Los *círculos* sobrepuestos indican elementos comunes.

vértice Punto donde se unen dos *semirrectas* formando un *ángulo*.

weight A measurement that tells how heavy an object is.

word form/word notation The form of a number that uses written words.

peso Medida que indica cuán pesado o liviano es un cuerpo.

forma verbal/notación verbal Forma de expresar un número usando palabras escritas.

Yy

yard (yd) A *customary* unit of *length* equal to 3 feet or 36 inches.

yarda (yd) Unidad *usual* de *longitud* igual a 3 pies o 36 pulgadas.

Zz

Zero Property of Multiplication
The property that states any number *multiplied* by zero is zero.

$$0 \times 5 = 0 \qquad 5 \times 0 = 0$$

propiedad del cero de la multiplicación
Propiedad que establece que cualquier número *multiplicado* por cero es igual a cero.

$$0 \times 5 = 0 \qquad 5 \times 0 = 0$$

Name _____

Work Mat 5: Tenths and Hundredths Models

Work Mat 5 Tenths and Hundredths Models WM1

Work Mat 6: Place-Value Chart

Ones			Decimals		
hundreds	tens	ones	tenths	hundredths	thousandths

Name

Work Mat 7: Centimeter Grid

Work Mat 7 Centimeter Grid WM3

Work Mat 8: Number Lines

WM4 **Work Mat 8** Number Lines